Malcolm Gerloch, Edwin C. Constable

# Transition Metal Chemistry

## The Valence Shell in d-Block Chemistry

**VCH**

Weinheim · New York
Basel · Cambridge · Tokyo

Dr. Malcolm Gerloch
University Chemistry Laboratory
Lensfield Road
Cambridge CB2 1EW
United Kingdom

Prof. Dr. Edwin C. Constable
Institut für Anorganische Chemie
Universität Basel
Spitalstr. 51
CH-4056 Basel
Switzerland

Published jointly by
VCH Verlagsgesellschaft, Weinheim (Federal Republic of Germany)
VCH Publishers, New York, NY (USA)


Editorial Director: Dr. Thomas Mager
Production Manager: Elke Littmann


Library of Congress Card No. applied for.


A catalogue record for this book is available from the British Library.


Die Deutsche Bibliothek – CIP-Einheitsaufnahme:
Gerloch, Malcolm:
Transition metal chemistry: the valence shell in d-block
chemistry / Malcolm Gerloch; Edwin C. Constable. -
Weinheim; New York; Basel; Cambridge; Tokyo: VCH, 1994
ISBN 3-527-29219-5 (Weinheim ...) brosch.
ISBN 3-527-29218-7 (Weinheim ...) Gb.
ISBN 1-56081-884-0 (New York)
NE: Constable, Edwin C.:

Composition: Hagedornsatz GmbH, D-68519 Viernheim. Printing: betz-druck gmbh, D-64291 Darmstadt.
Bookbinding: Industrie- und Verlagsbuchbinderei Heppenheim GmbH, D-64630 Heppenheim.
Printed in the Federal Republic of Germany

To our wives,
Gwynneth Neal-Freeman and Catherine Housecroft,
for their love, patience, hard work and cajolary.
Also to Satin, Index, Philby and Isis

# Preface

Transition metals comprise roughly half of the periodic table of elements. Their known chemistry occupies a rather larger fraction of non-carbon research literature to make up an enormous subject which continues to grow at a fast rate. No encyclopedia can encompass the century or more of achievement, let alone a single book. Here, we do not even begin to try. What is offered, however, is an outline of a theoretical structure for transition-metal chemistry at an elementary level that hopefully provides a consistent viewpoint of this widely varying and fascinating subject. By 'elementary' we mean early-to-mid UK degree level, and essentially non-mathematical: we do not mean, on the other hand, unsubtle, lacking in provocation or patronizing.

It has often been asserted that the 'driving forces' of inorganic chemistry vary throughout the periodic table so that we must focus on A here but on B there. If by this is meant that the *major* factors are A and B here and there, we have no quarrel. It is, however, utterly unsatisfactory for anyone coming to grips with the subject not to understand *why* A rules here and not there. We need an underlying structure and understanding if we wish more than to apply given recipes: something between the recipes and the impossibility of deriving chemistry from quantum theory and fundamental particles. This is a tall order. The present offering is an attempt within just the transition-metal series. Although the last chapter relates to the lanthanide series, we are mainly concerned with the first transition series only.

A central theme in our approach, which we believe to be different from those of others, is to focus on the changing chemistry associated with higher, middle and lower oxidation state compounds. The chemical stability of radical species and open-shell Werner-type complexes, on the one hand, and the governance of the 18-electron rule, on the other, are presented as consequences of the changing nature of the valence shell in transition-metal species of different oxidation state.

A goodly part of any text on 'theoretical' inorganic chemistry necessarily includes an account of crystal- and ligand-field theories. Usually, however, these theories are presented as a self-contained discipline. Although they have certainly provided wonderful opportunities for the exercise of group theory and physics within the inorganic chemistry syllabus, the student of chemistry can well be forgiven for wondering what they actually have to say about chemistry. It is necessary to go quite far into the purely symmetry-based aspects of crystal-field theory if only to explain the number of bands that occur in the spectra of transition-metal species or the gross features of their magnetic properties. And we do so in this book also, although we do not take the space to cover these matters all the way to the end of a UK bachelor course. What we do focus on particularly, though, and what is often too lightly skipped over in many other texts, is the light thrown by the crystal- and ligand-field theories upon 'chemical bonding and structure in the transition block. This is an interactive enterprise in that it is equally important to understand why ligand-field theory should 'work' anyway.·It is also important – though subtle, so we only make a start on it – to appreciate the utterly different nature of ligand-field theory on the one hand, and of molecular-orbital theory on the other.

In all these discussions, we separate, as best we might, the effects of the *d* electrons upon the bonding electrons from the effects of the bonding electrons upon the *d* electrons. The latter takes us into crystal- and ligand-field theories, the former into the steric roles of *d* electrons and the geometries of transition-metal complexes. Both sides of the coin are relevant in the energetics of transition-metal chemistry, as is described in later chapters.

We have agonized somewhat over the title of this book. Although it might put some readers off, we stuck with it for it really summarizes the kernel of our approach. This is not a compendium of chemical syntheses or properties, but rather an attempt to bring together in a single yet non-simplistic way many important bonding and theoretical principles that hopefully make more sense of this wide and fascinating subject. We hope that the path we have plotted through this important area of inorganic chemistry will commend itself to other teachers. Our lecture courses at Cambridge broadly follow this scheme, many of the central ideas of which were first presented in an article in *Coordination Chemistry Reviews* (**99**, **1990**, p 199).

One of us (M.G.) thanks Professors Bill Hatfield, Tom Meyer and their colleagues at the University of North Carolina, Chapel Hill, NC, U.S.A. for their hospitality whilst much of this book was written.

# Contents

# 1 An Introduction to Transition-Metal Chemistry

## 1.1 What is a Transition Element?

The transition elements comprise groups 3 to 12 and are found in the central region of the standard periodic table, an example of which is reproduced on the endpaper. This group is further subdivided into those of the first row (the elements scandium to zinc), the second row (the elements yttrium to cadmium) and the third row (the elements lanthanum to mercury). The term 'transition' arises from the elements' supposed *transitional* positions between the metallic elements of groups 1 and 2 and the predominantly non-metallic elements of groups 13 to 18. Nevertheless, the transition elements are also, and interchangeably, known as the transition metals in view of their typical metallic properties.

The chemistry of the transition elements has been investigated for two centuries, and in the past fifty years these elements and their compounds have proved to be a nearly ideal touchstone for many of the models which have been developed to understand structure and bonding. The elements range from the widespread to the extremely rare; iron is the fourth most abundant element (by weight) in the earth's crust, technetium does not occur naturally. Elements such as gold and silver have been known in the native state since antiquity, whereas technetium was first prepared in 1937. Most of the elements exhibit a typical silvery metallic appearance, but gold and copper are unique in their reddish coloration and mercury is the only metal which is liquid at ambient temperatures. Compounds of the transition elements account for the majority of coloured inorganic materials, and many pigments are relatively simple derivatives of these elements; however, not all transition-element compounds are coloured.

What are the common features that unite these elements? It is surprisingly difficult to find a single definition which satisfactorily encompasses all of the transition elements. The elements occur at that point in the periodic table where the $d$ orbitals are being filled. The first row transition elements coincide with the filling of the $3d$, the second row with the filling of the $4d$, and the third row with the filling of the $5d$ orbitals. We define a transition element as possessing filled or partially filled valence $d$ orbitals in one or more of its oxidation states. This definition excludes the elements in groups 13 to 18. The electron configurations of the transition elements are presented in Table 1-1.

The outer configurations of the transition metals in Table 1-1 imply, and detailed spectroscopic investigations confirm, that the $3d$ orbitals lie at higher energies than the $4s$ orbitals. On the other hand, the configurations of the $M^{2+}$ ions listed, in

**Table 1-1.** The electronic configurations of the transition elements.

| | | | | | |
|---|---|---|---|---|---|
| Scandium | [Ar]$3d^14s^2$ | Yttrium | [Kr]$4d^15s^2$ | Lanthanum | [Xe]$5d^16s^2$ |
| Titanium | [Ar]$3d^24s^2$ | Zirconium | [Kr]$4d^25s^2$ | Hafnium | [Xe]$4f^{14}5d^26s^2$ |
| Vanadium | [Ar]$3d^34s^2$ | Niobium | [Kr]$4d^45s^1$ | Tantalum | [Xe]$4f^{14}5d^36s^2$ |
| Chromium | [Ar]$3d^54s^1$ | Molybdenum | [Kr]$4d^55s^1$ | Tungsten | [Xe]$4f^{14}5d^46s^2$ |
| Manganese | [Ar]$3d^54s^2$ | Technetium | [Kr]$4d^55s^2$ | Rhenium | [Xe]$4f^{14}5d^56s^2$ |
| Iron | [Ar]$3d^64s^2$ | Ruthenium | [Kr]$4d^75s^1$ | Osmium | [Xe]$4f^{14}5d^66s^2$ |
| Cobalt | [Ar]$3d^74s^2$ | Rhodium | [Kr]$4d^85s^1$ | Iridium | [Xe]$4f^{14}5d^76s^2$ |
| Nickel | [Ar]$3d^84s^2$ | Palladium | [Kr]$4d^{10}5s^0$ | Platinum | [Xe]$4f^{14}5d^96s^1$ |
| Copper | [Ar]$3d^{10}4s^1$ | Silver | [Kr]$4d^{10}5s^1$ | Gold | [Xe]$4f^{14}5d^{10}6s^1$ |
| Zinc | [Ar]$3d^{10}4s^2$ | Cadmium | [Kr]$4d^{10}5s^2$ | Mercury | [Xe]$4f^{14}5d^{10}6s^2$ |

Table 1-2 for example, reveal the loss of electrons from the 4s shell in preference to the 3d, so that in these species the 4s orbitals are the higher in energy.

The explanation of these facts is not difficult but is subtle. We recall that the energies of all hydrogen orbitals belonging to the same principal quantum shell ($n$) are equal: the 3d, 3p and 3s hydrogen orbitals are degenerate. These orbital subsets

**Table 1-2.** The electronic configurations of the transition-metal ions in the divalent and trivalent states.

| | $M^{2+}$ | $M^{3+}$ | | | $M^{2+}$ | $M^{3+}$ |
|---|---|---|---|---|---|---|
| Scandium | | [Ar]$3d^04s^0$ | | Iron | [Ar]$3d^64s^0$ | [Ar]$3d^54s^0$ |
| Titanium | | [Ar]$3d^14s^0$ | | Cobalt | [Ar]$3d^74s^0$ | [Ar]$3d^64s^0$ |
| Vanadium | [Ar]$3d^34s^0$ | [Ar]$3d^24s^0$ | | Nickel | [Ar]$3d^84s^0$ | [Ar]$3d^74s^0$ |
| Chromium | [Ar]$3d^44s^0$ | [Ar]$3d^34s^0$ | | Copper | [Ar]$3d^94s^0$ | [Ar]$3d^84s^0$ |
| Manganese | [Ar]$3d^54s^0$ | [Ar]$3d^44s^0$ | | Zinc | [Ar]$3d^{10}4s^0$ | |

lose their degeneracy, however, in many-electron atoms. Orbitals with smaller orbital angular momentum quantum numbers (smaller $l$) possess increasing numbers of nodes in their radial functions and are referred to as increasingly 'penetrating'. Thus, a 3s electron experiences a larger effective nuclear charge and is more tightly bound than a 3p electron; a 3p is in turn more tightly bound than a 3d. Next, we recall that the energy separations between adjacent principal quantum shells in hydrogen decrease with increasing $n$. Taking both factors together, we expect that sooner or later, with respect to increasing atomic number, the more tightly bound orbital subsets of the $n^{th}$ principal quantum shell will be more tightly bound and decrease in energy below the higher orbital subsets of the $(n-1)^{th}$ principal shell.

For neutral atoms, that cross-over begins around the start of the transition-metal series. The balance between the $4s$ and $3d$ orbital energies is delicate, however, and other factors, not discussed so far, can reverse the general trend. One such factor is the exchange stabilization associated with the filled and half-filled $d$ shell. This will be familiar from discussion of ionization energies throughout the first long row of the periodic table when one considers the marked discontinuities at the $p^3$ and $p^6$ configurations; this theme is taken up in more detail in Chapter 8.

Now consider the ionization process yielding the $M^{2+}$ ions in the first row transition-metal series. The configuration adopted in the ion does not depend solely upon the relative orbital energies of the (energetically close) $4s$ and $3d$ orbitals in the neutral atom. It also depends upon the relative energies of the putative ions $3d^{n-2}4s^2$ and $3d^n4s^0$, for example. Let us consider each in turn. Removal of electrons from the $3d$ shell relieves some electron–electron repulsion and deshields the $4s$ orbital somewhat: both $3d$ and $4s$ shells will be more tightly bound in an $M^{2+}$ ion. Removal of electrons from the $4s$ shell, however, depletes the inner (sub-nodal) regions of their electron density with the result that the $3d$ orbitals are very much less well shielded and become much more tightly bound. It is perfectly possible in principle, and actually the case in practice, that the $3d$ orbital energy dips down below that of the $4s$ orbital as a result.

## 1.2 Complexes and Coordination Compounds

The systematic investigation of the chemistry of the transition elements began in the nineteenth century, and it rapidly became apparent that many of the compounds were somewhat different from those with which chemists were then familiar. There was a clear difference between the behaviour of simple ionic compounds such as sodium chloride and typical transition-element compounds such as $FeCl_2 \cdot 4H_2O$. It was also obvious that the compounds did not resemble the typically covalent compounds of organic chemistry. It was considered that many of the compounds formed by transition metals were of a complex constitution, and they were accordingly known as *complexes*.

The seminal studies on these complex compounds were conducted by Alfred Werner in an intensive period of work at the turn of the century.[*] A typical example of the problems that Werner addressed lies in the various compounds which can be obtained containing cobalt, ammonia and chlorine. Stable and chemically distinct materials with formulations $Co(NH_3)_nCl_3$ ($n = 4, 5$ or 6) can be isolated. The concepts of valency and three-dimensional structure in carbon chemistry were being developed at that time, but it was apparent that the same rules could not apply to

---

[*] Alfred Werner (1866 – 1919) was awarded the Nobel prize for chemistry as a recognition of these studies in 1913.

these complex compounds. Werner's key postulate was that a given metal ion could exert *two* different types of valence. The first of these related to the number of anionic groups which was associated with the compound and was termed the *primary valence*. Thus, the three compounds discussed above all contain three chloride groups and possess a primary valence of three. In modern terms, we would equate the Werner primary valency with the *oxidation state*. The novel idea that Werner introduced was that of *secondary valency*, which referred to the *number* of groups attached to a metal centre. The crucial observation was that the secondary valence could refer to the attachment of both anionic and *neutral* groups to the metal centre. Werner also recognized that in the same way that a metal had one or more characteristic primary valences, a given metal ion also had a number of characteristic secondary valences. He noted that the most common secondary valences were four and six. The secondary valence related to the number of groups which were directly attached to the metal atom in the *first* or *inner* sphere. Additional groups could be associated less strongly with a more distant *second* or *outer* sphere. Neutral ligands could occupy the inner but not the outer sphere. The cobalt centres in the three compounds $Co(NH_3)_nCl_3$ ($n$ = 4,5 or 6) all possess a primary valence (oxidation state) of three, and the characteristic secondary valence for cobalt(III) is six. Thus $Co(NH_3)_4Cl_3$ possesses four ammonias and two chlorides in the inner sphere and a chloride in the outer sphere, $Co(NH_3)_5Cl_3$ possesses five ammonias and one chloride in the inner sphere and two chlorides in the outer sphere, and $Co(NH_3)_6Cl_3$ possesses six ammonias in the inner sphere and three chlorides in the outer sphere. Chemical and physical evidence was presented to support these contentions.[*] At the time Werner developed a number of descriptions for the bonding in such compounds which were related to the structures of more familiar organic species. We will not be concerned with these, but note that secondary valence is equivalent to the modern term *coordination number*.

The interactions in such compounds are now better understood, and the term complex now has a more specific meaning. Not all transition-metal compounds are complexes, but many are. The terms complex and *coordination compound* are now used almost interchangably.

---

[*] Particular use was made of conductivity measurements of cobalt(III) and platinum(II) complexes which allowed a facile determination of the number and type of ions present in solution. For example, the compounds $Co(NH_3)_nCl_3$ would give a monocation and an monoanion ($n$=4), a dication and two monoanions ($n$ = 5) and a trication and three monoanions ($n$=6) respectively. In some cases, it was also possible to distinguish *chemically* between inner and outer sphere chloride by precipitation of the outer sphere species as AgCl.

## 1.3  The Coordinate Bond

In a typical covalent bond, such as is found between carbon and hydrogen in methane, each atom is considered to contribute one electron to the two-electron, two-centre bond which is formed. However, we can envisage a second type of covalent bond in which we still have a two-centre, two-electron bond, but where both of the electrons come from *one* of the atoms or from a molecule. This type of bond is known variously as a *coordination*, a *dative covalent* or a *donor–acceptor* bond. A compound containing such bonding is known as a *coordination compound*. The atom (or molecule) which provides the two electrons is known as the *donor*. The other atom (or molecule) is known as the *acceptor*. The term *complex* is used to describe a coordination compound in which the acceptor is a metal (usually, but not necessarily, a transition metal) atom or ion. In those coordination compounds in which the acceptor is a metal atom or ion, the donor is known as a *ligand* (from the Latin word *ligare*, which means to bind). It is interactions of this nature which are responsible for the binding of ligands to a metal ion and with which we will be concerned for the remainder of this book. Note that this is a *formal* description of the donor–acceptor interaction between the ligand and the metal and conveys little about the *actual* electron distribution. It is in no way a comment about the 'real' electron distribution in transition-metal compounds. We will return to this topic in Section 1.8.

## 1.4  Ligand Types

It is probably true that almost every conceivable molecule, atom or ion could act as a ligand under some circumstance or other. However, certain types of ligands are commonly encountered, and it is these, together with the vocabulary which they generate, that we introduce at this stage.

The majority of ligands are either neutral or anionic. Those which coordinate to a metal ion through a single atom are described as monodentate or unidentate. Examples of such ligands which we have encountered thus far include water, ammonia and chloride. A more extensive listing of common ligands is found in Table 1-3. We stress at this point that there is no difference in kind between the interactions of a metal centre with either neutral or anionic ligands.

A number of general features in Table 1-3 is apparent. Complexes may be cationic, neutral or anionic. Ligands may be simple monatomic ions, or larger molecules or ions. Many ligands are found as related neutral and anionic species (for example, water, hydroxide and oxide). Complexes may contain all of the same type of ligand, in which case they are termed *homoleptic*, or they may contain a variety of ligand types, whereby they are described as *heteroleptic*. Some ligands such as nitrite or thiocyanate can coordinate to a metal ion in more than one way. This is described as *ambidentate* behaviour. In such cases, we commonly indicate

**Table 1-3.** Some typical monodentate ligands and representative complexes that they form.

| Monodentate ligands | Donor atom | Example |
| --- | --- | --- |
| $O^{2-}$ | O | $[MnO_4]^-$ |
| Halides, $F^-$, $Cl^-$, $Br^-$, $I^-$ | F, Cl, Br, I | $[NiCl_4]^{2-}$, $[CrF_6]^{3-}$ |
| $H^-$ | H | $[ReH_9]^{2-}$ |
| $NCS^-$ | N or S | $[Cr(NH_3)_5(NCS)]^{2+}$ |
|  |  | $[Cr(NH_3)_5(NCS)]^{2+}$ |
| $NO_2^-$ | N or O | $[Co(NH_3)_5(NO_2)]^{2+}$ |
|  |  | $[Co(NH_3)_5(ONO)]^{2+}$ |
| $RS^-$ | S | $[Fe(SPh)_4]^-$ |
| $CN^-$ | C | $[Fe(CN)_6]^{3-}$ |
| $HO^-$ | O | $[Zn(OH)_4]^{2-}$ |
| $H_2O$ | O | $[Mn(H_2O)_6]^{2+}$ |
| $NH_3$ | N | $[Co(NH_3)_6]^{3+}$ |
| phosphines, $PR_3$ | P | $[Pt(PMe_3)_4]$ |
| pyridine, py | N | $[Ni(py)_6]^{2+}$ |
| CO | C | $[Mn(CO)_5]^-$ |
| RCN | N | $[Ru(NH_3)_5(NCMe)]^{2+}$ |
| $(CH_3)_2S$ | S | $[Pt\{(CH_3)_2S\}_2Cl_2]$ |

the atom which is involved in coordination to the metal by italicizing it, as in the *N*–bonded thiocyanate in the ion $[Cr(NH_3)_5(NCS)]^{2+}$.

Ligands which interact with a metal ion through two or more donor atoms are of particular importance in coordination chemistry. The number of donor atoms involved is indicated by the *denticity* – a didentate (or bidentate) ligand interacts with metals through two donor atoms, a tridentate (or terdentate) through three, and so on. If two or more of the donor atoms are interacting with the same metal centre, the ligands are described as *chelating* and the complexes as *chelates*. It is generally found that there is an extra stability associated with complexes which contain chelating ligands – the so-called *chelate effect* (this is discussed in detail in Chapter 9). In Table 1-4 we list some common polydentate ligands together with the abbreviations by which they are commonly known. Once again, note that both neutral and anionic ligands are found, and that the range of donor atoms is great. A new feature of these polydentate ligands is that they may contain mixtures of different donor atoms within the same ligand. Note also that a range of cyclic ligands is known, each of which provides a central cavity for a metal ion. The study of such macrocyclic or encapsulating ligands is of considerable current interest.

**Table 1-4.** Some typical polydentate ligands and their complexes.

| Polydentate ligands | Donor atoms | Example |
|---|---|---|
| *Didentate* | | |
| acetylacetonate, pentane-2,4-dionate, acac | O,O' | $[Cr(acac)_3]$ |
| oxalato, ox | O,O' | $[Fe(ox)_3]^{3-}$ |
| glycinato, gly | N,O | $[Cu(gly)_2]$ |
| $H_2NCH_2CH_2NH_2$ 1,2-diaminoethane, ethylenediamine, en. | N,N' | $[Co(en)_2]^{3+}$ |
| $Ph_2PCH_2CH_2PPh_2$ bis(diphenylphosphino)-ethane, dppe | P,P' | $[Fe(dppe)_2(CO)]$ |
| 1,2-bis(dimethylarsino)-benzene, diars | As,As' | $[CrCl_4(diars)]^-$ |
| | S,S' | $[Re(S_2C_2Ph_2)_3]$ |
| 2,2'-bipyridine, 2,2'-dipyridyl, bipy, bpy | N,N' | $[Mn(bpy)_3]^{2+}$ |
| 1,10-phenanthroline, phen | N,N' | $[Ru(phen)_3]^{2+}$ |

**Table 1-4.** (Continued)

| Polydentate ligands | Donor atoms | Example |
|---|---|---|
| *Tridentate* | | |
| $H_2NCH_2CH_2NHCH_2CH_2NH_2$ Diethylenetriamine, 1,4,7-triazaheptane, bis(2-aminoethyl)amine, dien. | N,N',N" | $[Co(dien)_2]^{3+}$ |
| 2,2':6',2"-terpyridine, terpyridyl, tpy, terpy | N,N',N" | $[Cr(tpy)_2]^{3+}$ |
| *Macrocyclic* | | |
| 18-crown-6 | $O_6$ | $[K(18\text{-crown-}6)]^+$ |
| 1,4,8,11-tetrazacyclo- tetradecane, cyclam | N,N',N",N''' | $[Ni(cyclam)]^{2+}$ |

## 1.5  Coordination Number

The coordination number of a metal ion in a complex is defined as the number of donor atoms bonded to the metal centre. In most cases it is simple to determine. The coordination number is six in the complex species $[Fe(H_2O)_6]^{2+}$, $[Fe(py)_6]^{2+}$, $[Fe(CN)_6]^{4-}$, $[Fe(bpy)_3]^{2+}$ and $[Fe(tpy)_2]^{2+}$. Note that when chelating ligands are involved, it is the number of *donor atoms* and not the number of ligands which defines the coordination number. The coordination number is not so easily defined when we consider those organometallic complexes in which ligation involves $\pi$-bonding of two or more centres within the ligand to a metal. For example, in the anion $[PtCl_3(H_2C=CH_2)]^-$, the platinum interacts equally with the *two* carbon atoms of the ethene ligands. Is the coordination number four or five? A special nomenclature has been developed to describe the types of interaction encountered in organometallic compounds, and the concept of coordination number is probably not particularly useful in this context.

Again, remember that coordination number is equivalent to Werner's secondary valence.


# 1.6  Geometrical Types and Isomers


Coordination compounds show a wide variety of regular, and an infinite range of irregular, geometries for the arrangement of the ligands about the metal centre. However, for the first row transition metals, a few geometries by far outweigh all of the others. The regular polyhedra upon which complexes are commonly based are the octahedron (six coordination) and the tetrahedron (four coordination). A significant number of four coordinate complexes exhibit a planar geometry and in Chapter 7 we rationalize the occurrence of this structural geometry. One of the consequences of complexes adopting specific geometries is the occurrence of isomers. We review these only briefly, and the interested reader will find more information in the "suggestions for further reading" at the close of this chapter.

Several different types of isomers arise in transition-metal coordination compounds, and these are described below.

*Structural isomers*: These are compounds in which the isomers are related by the interchange of ligands inside the coordination sphere for those outside it. A classical example of this phenomenon is observed in the compounds of formula $CrCl_3(H_2O)_6$. As usually obtained from chemical suppliers, this is a green solid in which only two of the chloride ions are coordinated to the metal. This is formulated $[Cr(H_2O)_4Cl_2]Cl \cdot 2H_2O$. Solutions of this compound in water slowly turn blue-green as a coordinated chloride ion is replaced by a water molecule and the complex $[Cr(H_2O)_5Cl]Cl_2 \cdot H_2O$ may be isolated. More commonly, structural isomers are related by the exchange of *anionic* ligands and counter ions, rather than neutral ligands. Typical examples include the pair of complexes $[Co(en)_2Br_2]Cl$ and $[Co(en)_2BrCl]Br$.

*Linkage isomerism*: This is a special type of structural isomerism in which the differences arise from a particular ligand which may coordinate to a metal ion in more than one way. In Table 1-3 we indicated that a ligand such as thiocyanate could bond to a metal through either the nitrogen or the sulfur atom, and the complex ions $[Co(NH_3)_5(NCS)]^{2+}$ and $[Co(NH_3)_5(SCN)]^{2+}$ are related as linkage isomers.

*Coordination isomerism*: This is an interesting type of isomerism which can occur with salts in which both the cation and the anion are complex ions. Consider the salt $[Co(bpy)_3][Fe(CN)_6]$ containing one cobalt (III) and one iron (III) centre: coordination isomers of this would include $[Fe(bpy)_3][Co(CN)_6]$, $[Co(bpy)_2(CN)_2][Fe(bpy)(CN)_4]$, $[Fe(bpy)_2(CN)_2][Co(bpy)(CN)_4]$, and $[Co(bpy)_3][Fe(CN)_6]$.

*Geometrical isomerism*: This is an important topic which played a crucial role in the development of coordination chemistry. Werner used the number of isomers which could be isolated for a range of cobalt(III) complexes to establish the octahedral character of the $CoL_6$ species.

A planar complex of the type $[Pt(NH_3)_2Cl_2]$ can exist in two forms depending upon the relative spatial orientation of the two chloride ligands. They can be at 90° to each other to give the *cis* form (**1.1**), or at 180° to give the *trans* isomer (**1.2**).

**1.1**                                **1.2**

In six coordinate complex ions such as $[Co(NH_3)_4Br_2]^+$, a similar situation exists, in which the bromine ligands adopt either a *cis* (**1.3**, **1.4**) or a *trans* arrangement (**1.5**). The reader should note the identity of the *cis* isomers despite the different drawings (**1.3** and **1.4**). In a similar manner, complexes of the type $[MX_3Y_3]$ may adopt two structures, depending upon the relative arrangement of the three identical groups in the octahedron. If the three X groups are arranged about a single triangular face, then the *facial* (or *fac*) isomer (**1.7**) is obtained, whereas if they are arranged in three of the four sites of the equatorial plane, the *meridional* (or *mer*) isomer (**1.6**) is obtained.

*cis*                        *trans*

**1.3**              **1.4**                **1.5**

*mer*                        *fac*

**1.6**                        **1.7**

Notice the 'loose' use of the term octahedral to describe six-coordinate complexes which are based upon an octahedral geometry, but which, by virtue of the presence of different ligand types, are of lower symmetry than $O_h$. This is a common usage which should give rise to no difficulties. Note also how introduction of chelating ligands into the coordination shell may reduce the number of isomers which are possible. Thus, although there are two isomers of [Pt(NH₃)₂Cl₂], it is only possible to form the *cis* isomer of [Pt(en)Cl₂] (**1.8**). This is because the relative positions of the nitrogen donor atoms in the en ligand are dictated by the $CH_2CH_2$ linker group – the two donor atoms cannot 'stretch' to occupy *trans* positions. Similarly, it is only possible to obtain the *cis* isomer of the cation $[Co(NH_3)_4(en)]^{3+}$ (**1.9**).

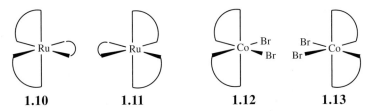

**1.8**          **1.9**

A final type of isomerism which we mention here also arises most commonly when chelating ligands are present. If a molecule possesses neither a plane nor a centre of symmetry, it is chiral. (This definition is not strictly correct, but will suffice for most transition-metal complexes.) Chiral species may exist in two forms which are related as mirror images. These have identical chemical and physical properties unless they are interacting with something else which is chiral, in which case they differ. That may be a chiral reagent (to give diastereomeric compounds) or polarized light. A typical example of a chiral complex is found when three chelating ligands are coordinated to an octahedral centre, as in the cation $[Ru(bpy)_3]^{2+}$. Two different forms of this cation, related as mirror images, are possible (**1.10** and **1.11**). These may be separated by formation of salts with chiral anions, and exhibit different and opposite rotations of polarized light. Note also that the cation $[Co(en)_2Br_2]^+$ (**1.12** and **1.13**) is chiral, but $[Co(NH_3)_4Br_2]^+$ is not.

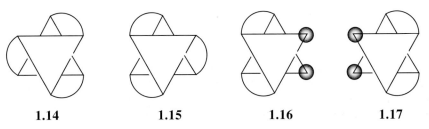

**1.10**          **1.11**          **1.12**          **1.13**

Another way of drawing these isomers emphasizes the three-fold nature of the basic octahedron rather than its four-fold properties (**1.14 – 1.17**).

**1.14**          **1.15**          **1.16**          **1.17**

## 1.7  Oxidation State

Oxidation state is a frequently used (and indeed misused) concept which apportions charges and electrons within complex molecules and ions. We stress that oxidation state is a *formal* concept, rather than an accurate statement of the charge distributions within compounds. The oxidation state of a metal is defined as the formal charge which would be placed upon that metal in a purely ionic description. For example, the metals in the gas phase ions $Mn^{3+}$ and $Cu^+$ are assigned oxidation states of +3 and +1 respectively. These are usually denoted by placing the formal oxidation state in Roman numerals in parentheses after the element name; the ions $Mn^{3+}$ and $Cu^+$ are examples of manganese(III) and copper(I).

---

**Box 1-1**

Older texts often employ an alternative nomenclature in which the suffixes -ous and -ic are encountered. In general, these labels only apply to the most common oxidation states of the metals, -ic referring to the higher oxidation state and -ous to the lower. Using this nomenclature, copper(II) is referred to as cupric and copper(I) as cuprous. The system works well if there are only two common oxidation states for a metal ion, but if there are more, the scheme becomes either ambiguous or unwieldy as a variety of prefixes are added.

---

It is usually easy to define the oxidation state for *simple* compounds of the transition metals. In the case of neutral compounds, we assign charges as if the compound were ionic. Thus, $MnCl_2$ is regarded as $\{Mn^{2+}, 2Cl^-\}$ and is correctly described as manganese(II) chloride. Similarly, $WO_3$ as $\{W^{6+}, 3O^{2-}\}$ is tungsten(VI) oxide. Since ligands which bear no formal charges in an ionic formulation may be ignored, $[Cr(H_2O)_3Cl_3]$ is a chromium(III) compound, and $Ni(OH)_2$, $NiBr_2$, $NiBr_2 \cdot 3H_2O$, $NiBr_2 \cdot 6H_2O$ and $NiBr_2 \cdot 9H_2O$ are all nickel(II) compounds. The assignment of oxidation state makes no implications regarding the nature of the bonding within the molecule – all of the various hydrated forms of $CrCl_3$ are chromium(III) compounds. Oxidation state is merely a formal scheme: there is no implication that tungsten(VI) oxide necessarily contains $W^{6+}$ ions. Furthermore, problems with the assignment of oxidation state can arise with even apparently simple compounds. Consider, for example, $Fe_3O_4$. If the compound were ionic, we would have four $O^{2-}$ ions. In order for the entire compound to be neutral, the three iron atoms must possess an overall charge of +8. The ensuing assignment of an oxidation state of +8/3 to each iron is not particularly meaningful. A compound of this type is best regarded as a mixed oxidation state oxide, $\{FeO + Fe_2O_3\}$ or $Fe^{II}Fe_2^{III}O_4$, in which there are both iron(II) and iron(III) centres.

Cations and anions are treated in an exactly similar manner, remembering to take the overall charge of the species into account. If only neutral ligands are present, the oxidation state of the metal ion is equal to the overall charge on the ion. Thus, $[Fe(H_2O)_6]^{3+}$ and $[Ni(NH_3)_6]^{2+}$ are iron(III) and nickel(II) complexes respectively. If charged ligands are present, formal charges are assigned on the basis of an ionic description. Thus, the ion $[Ni(CN)_4]^{2-}$ is treated as containing a cationic nickel centre and four anionic cyanides. Since the four cyanides give a total charge of $-4$, the nickel must be assigned a charge of $+2$ in order for the ion to possess an overall charge of $-2$, and it is therefore a nickel(II) complex. Similarly, $[MnO_4]^-$ is treated as $\{Mn^{7+}, 4O^{2-}\}$ and is a manganese(VII) compound. Once again, we stress that this in no way implies that the ion $[MnO_4]^-$ actually contains a $Mn^{7+}$ ion. By the way, aqueous solutions of transition-metal compounds frequently contain ions such as $[M(H_2O)_6]^{n+}$: as water is the most common solvent encountered in chemical reactions, these species are often (but incorrectly) referred to as solutions containing $M^{n+}$ ions (see Box 1-2).

It is quite possible for a metal centre to possess a zero or negative oxidation state. Thus, the species $[Cr(CO)_6]$ and $[Fe(CO)_4]^{2-}$ are chromium(0) and iron($-2$) complexes. We will see in a later chapter that it is not a coincidence that these low formal oxidation states are associated with ligands such as carbon monoxide.

Some ligands pose problems in the assignment of a formal oxidation state to a metal centre. Nitric oxide is a case in point. The ligand may be formulated as either anionic $NO^-$ or cationic $NO^+$, and there follows the appropriate ambiguity in assignment of the oxidation state of the metal ion to which it is bonded. These problems arise when it is not clear as to what charge is appropriate to assign to the ligands in the ionic limit. We have repeatedly emphasized the formal character of the concept of oxidation state and turn now to a different general concept which helps us address the *real* electron distributions in compounds.

---

**Box 1-2**

It is very common for inorganic chemists to 'neglect' or 'ignore' the presence of solvent molecules coordinated to a metal centre. In some cases, this is just carelessness, or laziness, as in the description of an aqueous solution of cobalt(II) nitrate as containing $Co^{2+}$ ions. Except in very concentrated solutions, the actual solution species is $[Co(H_2O)_6]^{2+}$. In other cases, it is not always certain exactly what ligands remain coordinated to the metal ion in solution, or how many solvent molecules become coordinated. Solutions of iron(III) chloride in water contain a mixture of complex ions containing a variety of chloride, water, hydroxide and oxide ligands.

When dealing with the kinetic or thermodynamic behaviour of transition-metal systems, square brackets are used to denote concentrations of solution species. In the interests of simplicity, solvent molecules are frequently omitted (as are the square brackets around complex species). The reaction (1.1) is frequently written as equation (1.2).

$$[Co(H_2O)_6]^{2+} + 4Cl^- = [CoCl_4]^{2-} + 6H_2O \qquad (1.1)$$

$$Co^{2+} + 4Cl^- = [CoCl_4]^{2-} \qquad (1.2)$$

Whilst this will be satisfactory when dealing with kinetic data in which reactions involving the solvent will not explicitly appear in the rate equations, it is not appropriate when we consider equilibrium constants. As an exercise, consider the formation of $[Ni(en)_3]^{2+}$ from aqueous solutions of nickel(II) chloride and en (en $= H_2NCH_2CH_2NH_2$); write the equations with the inclusion and the omission of the water molecules. Can you recognize the driving force for the formation of the chelate in each case?

## 1.8  Electroneutrality Principle

It was recognized early on that the formality of 'dative covalency' in coordination compounds presents some difficulties. Many inorganic compounds are conveniently thought of as ionic salts in which there is an essentially complete charge separation between cationic and anionic species. Compounds of the group 1 and group 2 metals tend to be readily categorized as 'ionic'. Dissolution of a salt of a group 1 or a group 2 metal results in the formation of solutions containing solvated cations and anions – only in concentrated solutions are there significant cation–anion interactions beyond simple ion-pairing. However, the ionic model does not appear to be suitable for the description of the properties of many transition-metal compounds. For example, the compound $K_4[Fe(CN)_6]$ dissolves in water to give solutions containing solvated potassium ions and the $[Fe(CN)_6]^{4-}$ ion, rather than solvated potassium, iron(II) and cyanide ions. The interactions between the cyanide and the iron(II) centre appear to result in longer lived species than result from simple electrostatic interactions of the type observed in sodium chloride. This is, of course, the sort of argument which led to the development of the description of coordination compounds in terms of donor–acceptor interactions between the ligands and the metal centre.

Let us now examine the consequences of the formation of a donor–acceptor bond in a little more detail. If the donor–acceptor bond is completely covalent, then we record net transfer of one unit of charge from the donor to the acceptor as a direct consequence of the equal sharing of the electron pair between the two centres. This result leaves a positive charge on the donor atom and a negative charge on the acceptor atom. The limiting 'ionic' and 'covalent' descriptions of a complex cation such as $[Fe(H_2O)_6]^{3+}$ are shown in Fig. 1-1.

**Figure 1-1.** Limiting valence bond representations of the cation $[Fe(H_2O)_6]^{3+}$.

We have already commented that the 'ionic' structures are not in accord with the chemical properties of coordination compounds. However, there are also a number of objections to the covalent description. The charge distribution is such that the iron(III) centre bears a three minus charge, whereas the oxygen atoms of the water ligands each bear a single positive charge. This would be unrealistic in view of the electronegativities of these elements (Fe, 1.8; O, 3.5), which predict that the Fe–O bond should be polarized in the sense $Fe^{\delta+}-O^{\delta-}$. The problem was addressed by Pauling, who recognized that, in reality, it was not appropriate to describe most bonds as being purely 'ionic' or purely 'covalent'. He developed a description of the bonding between a metal and its ligands which included considerable ionic character in the metal – ligand bonds within a basically covalent régime. In order to determine the amount of 'ionic' character within a given metal – ligand bond, Pauling framed his *electroneutrality principle*. In this, he opined that the actual distribution of charges within a molecule is such that charges on any single atom are within the range – 1 to +1.

We shall see how this works by reconsidering the ion $[Fe(H_2O)_6]^{3+}$. An 'ionic' description results in a +3 charge on the metal centre, whilst a 'covalent' description gives the metal a $-3$ charge. Now the electroneutrality principle suggests that the 'ideal' charge on the metal centre is zero. That would be achieved if the iron centre gains a total of three electrons from the six oxygen donor atoms; in other words, if each oxygen loses one half of an electron rather than the whole electron which the fully covalent model demands (**1.18**). Pauling describes this situation as 50% covalent (or 50% ionic).

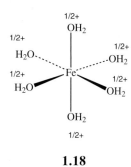

**1.18**

We shall return to this topic in Chapter 9. Remember that the unequal distribution of electrons within bonds results in a continuous variation from 'covalent' to 'ionic' bonding.

## 1.9  Rationalization of Complex Geometries

The coordination geometries of main group compounds are generally rationalized in terms of the Valence Shell Electron Pair Repulsion (VSEPR) or Gillespie-Nyholm model. The reader will recall that in this scheme, the spatial arrangement of atoms and groups about a central atom is dictated solely by the number of such groups and by the number of stereochemically active lone pairs. The model only considers mutually repulsive interactions between the various ligands and lone pairs present in the valence shell (or equivalently, between bond pairs and lone pairs in the valence shell) and makes no assumptions about the nature of the bonding except insofar as it is predicated upon a particular number of lone pairs. The assumption is generally made that all electrons in the valence shell – both lone pairs and bonding pairs – are stereochemically active. This model is remarkable for both its simplicity and its general applicability. The basic method involves the totalling of the number of atoms or groups and lone pairs associated with the central atom, and deriving a structure based upon the appropriate *n*-vertex polyhedron.

---

**Box 1-3**

These are the polyhedra which are used for the basic structural types:

| vertices | polyhedron | coordination geometry |
|---|---|---|
| 3 | triangle | trigonal planar |
| 4 | tetrahedron | tetrahedral |
| 5 | trigonal bipyramid | trigonal bipyramidal |
| *or* | square-based pyramid | square (-based) pyramidal |
| 6 | octahedron | octahedral |

---

If the central atom has different groups or atoms around it, or if one or more of the vertices of the polyhedron is occupied by a lone pair, then variations in bond angles will occur such that distorted polyhedral arrangements are obtained. In its quantitative forms, the VSEPR model parameterizes each individual interaction and makes very accurate predictions of the distortions which are to be expected.

This model has been successfully applied to the structures of many thousands of main group compounds, and bond angles within a few degrees of the experimentally observed values are usually correctly predicted. The basic model is only concerned with *repulsive* interactions between atoms and other atoms or lone pairs. Nowhere in the model is any consideration given to the attractive nature of the bonding which is present (single, double or triple bonds) or to the ways in which the central atom atomic orbitals must be utilized in attaining the desired geometry: once again, except for the assignment of the number of nonbonding electron pairs no assumptions about the bonding are made. Many texts suggest that the VSEPR model may not be usefully applied to transition-metal compounds. This is not so.

Kepert has developed a repulsion model for the prediction and rationalization of angular coordination geometry in transition-metal complexes at various levels of detail. His basic model considers the ligands to be arranged upon the surface of a sphere enclosing the central metal ion. The distances between donor atoms of chelating ligands are fixed as determined by intra-ligand bonding. This apart, Kepert's model allows for free variation in the angular geometry, that is, free movement of all donor atoms (or chelate groups as appropriate) on the surface of the notional sphere, subject to a $1/r^n$ repulsive force between them. Nowhere in the basic model is any consideration given to the nature of the metal–ligand bonding or to the steric potential of the $d$ configuration. In application to thousands of compounds, the model consistently predicts angular geometries which agree with experiment to within about $2°$. The relative energies of these conformational minima depend, of course, upon the value of $n$ in the repulsion law invoked ($n = 2$, 6 and 12 have all been investigated) but the angular *positions* of these minima are almost independent of $n$.

There is one striking group of exceptions to the otherwise almost unbroken success of Kepert's approach. No model predicated solely upon the repulsions between monodentate ligands (or between bonds) can account for the planarity of some four-coordinate complexes. Yet hundreds of planar $d^8$ complexes like $[Ni(CN)_4]^{2-}$ or $[PtCl_4]^{2-}$ are known. Clearly, Kepert's model is to be augmented and we discuss this matter further in Chapter 7.

---

**Box 1-4**

The compound $[Zn(tpy)Cl_2]$ (**1.19**) contains a planar tridentate ligand with nitrogen donor atoms. The geometry is often described as trigonal bipyramidal with the three 2,2':6',2"-terpyridine donor atoms occupying one equatorial and the two axial sites. Kepert's calculations actually predict a geometry that is far closer to the ideal square-based pyramid. His predictions are well confirmed by crystallographic analysis.

**1.19**

---

Thus far, we have only considered the *angular* geometry of complexes; variations in bond lengths also pose challenges. For example, the gross inequality of bond lengths in $[NiF_6]^{3-}$ and many copper(II) and chromium(III) complexes requires an explanation. Questions of this kind are also addressed in Chapter 7.

# 1.10 Review of Properties of Transition-Metal Compounds

Finally, we summarize some of the properties of transition-metal compounds and attempt to distinguish those which are characteristic of a *transition-metal complex* as opposed to *any metal complex*.

*Variable oxidation state* – One obvious feature of transition-metal chemistry is the occurrence of a number of characteristic oxidation states for a particular metal

**Table 1-5.** The oxidation states of first row transition-metals.

| | Sc | Ti | V | Cr | Mn | Fe | Co | Ni | Cu | Zn |
|---|---|---|---|---|---|---|---|---|---|---|
| 0 and lower | | | √ | √ | √ | √ | √ | √ | √ | |
| +1 | | | √ | √ | √ | √ | √ | √ | √√ | |
| +2 | | √ | √ | √√ | √√ | √√ | √√ | √√ | √√ | √√ |
| +3 | √√ | √ | √√ | √√ | √ | √√ | √√ | √ | √ | |
| +4 | | √√ | √√ | √ | √ | √ | √ | √ | | |
| +5 | | | √√ | √ | √ | √ | √ | | | |
| +6 | | | | √√ | √ | √ | | | | |
| +7 | | | | | √√ | | | | | |

√    Known
√√   Commonest oxidation states

ion. In general, these oxidation states are readily interconverted. This tendency to form a variety of oxidation states is displayed in Table 1-5.

Note that the occurrence of a maximum oxidation state, corresponding to the removal of all the valence shell electrons and the adoption of a $d^0$ configuration, does not occur after manganese. In Chapter 9 we see how this reflects the contraction of the poorly penetrating $3d$ orbitals as the nuclear charge increases and it becomes progressively more difficult to remove electrons.

The exhibition of variable valency is indeed a characteristic of transition metals. Main group metal ions such as those of groups 1 or 2 exhibit a single valence state. Other main group metals may show a number of valencies (usually two) which are related by a change in oxidation state of two units. This is typified by the occurrence of lead(IV) and lead(II) or thallium(III) and thallium(I). However, all the transition metals exhibit a range of valencies that is generally not limited in this manner.

*Low oxidation states* – An important characteristic of transition metal chemistry is the formation of compounds with low (often zero or negative) oxidation states. This has little parallel outside the transition elements. Such complexes are frequently associated with ligands like carbon monoxide or alkenes. Compounds analogous to Fe(CO)$_5$, [Ni(cod)$_2$] (cod = 1,4-cyclooctadiene) or [Pt(PPh$_3$)$_3$] are very rarely encountered outside the transition-metal block. The study of the low oxidation compounds is included within organometallic chemistry. We comment about the nature of the bonding in such compounds in Chapter 6.

*Colour* – A striking feature of transition-metal compounds is their colour. Whether it is the pale blue or pink hues of copper(II) sulfate and cobalt(II) chloride, or the intense purple of potassium permanganate, these colours tend to be associated most commonly with transition-metal compounds. It is rare for compounds of main group metals to be highly coloured.

*Unpaired electrons and magnetism* – One of the consequences of the open (incompletely filled) $d^n$ configuration of transition-metal ions may be the presence of one or more unpaired electrons. Such compounds could be described as radicals, and they are detected by techniques such as electron spin resonance spectroscopy.

However, while transition-metal ions often contain unpaired electrons, they exhibit none of the reactivity that is commonly associated with such radicals outside the *d* block. There is no behaviour comparable to that of the highly reactive and short lived radicals such as $CH_3^{\cdot}$. Also associated with the presence of unpaired electrons in these species is the phenomenon of paramagnetism. The long – term stability of many compounds with unpaired electrons is a characteristic of the transition-metal series.

*Formation of coordination compounds and variable coordination number* – Both the transition and the main group metal ions form coordination compounds. There is no difference in kind between the complexes formed between cobalt(III) and ammonia and those between lithium and water. Though the absolute stabilities may vary, large ranges of stability constants are observed for both main group and transition-metal ions. Transition-metal complexes may gain or lose ligands to change geometry and so do main group complexes. The existence of coordination chemistry in the transition-metal block does not set these metals apart from those of the main groups.

# Suggestions for further reading

1. F. Basolo, R.C. Johnson, *Coordination Chemistry,* 2nd ed., Science Reviews.
   – This is an easy to read introduction to the area.
2. J.E. Huheey, E.A. Kieter and R.L. Kieter, *Inorganic Chemistry,* Harper Collins, New York, **1993.**
   – An excellent general introduction to inorganic chemistry, with first rate chapters dealing with transition metal chemistry.
3. F.A. Cotton, G. Wilkinson, *Advanced Inorganic Chemistry,* 5th ed., Wiley, New York, **1989.**
   – A relatively comprehensive work with a great deal of descriptive material concerned with transition metal chemistry.
4. A.G. Sharpe, *Inorganic Chemistry,* 2nd ed., Longman, London, **1992.**
   – A general text with a number of relevant chapters.

# 2 Focus on the $d^n$ Configuration

## 2.1 Spectral Features

We discover a far-reaching generality of transition-metal compounds simply by looking at bottles on the laboratory shelves. By and large – and with many exceptions to be sure – compounds of transition metals are coloured, while those of the main group metals are not. Furthermore, the colours are gentle rather than vivid – weak rather than strong – and often group together with the metal ion involved. Thus, many copper(II) complexes are blue, while those of nickel(II) are green; manganese(II) compounds are only weakly coloured; a wide range of colours are associated with the different oxidation states of vanadium. Look again and these generalizations are seen to fail, but there are clearly some patterns to be found. We shall expend considerable time and effort discovering and understanding these patterns and generalities, not just because it might be fun to make theories about the pretty colours but because they are the outward manifestations of much of the underlying electronic structure in transition-metal complexes. To be honest, it is only with hindsight that we can say what is probed by the spectral features, so that many parts of the arguments we shall develop must be by assertion: but then, that is true of other, more conventional, approaches too.

Electronic absorption spectra of a few typical transition-metal complexes are shown in Fig. 2-1. The following features are to be noted.

a. All absorptions are broad, often up to 2000 cm$^{-1}$ wide yet occasionally down to 100 cm$^{-1}$. Atomic line spectra are of the order 1 cm$^{-1}$ in width.

b. Most bands in the near IR, visible and near UV are weak and about $10^2$ to $10^4$ times weaker than bands characterizing dyestuffs. These are called '$d-d$' bands.

c. Often, much more intense bands occur at higher energies, usually in the ultraviolet region. These comprise so-called 'charge-transfer' bands as well as ligand-centred $n-\pi^*$ and $\pi-\pi^*$ transitions.

d. The spectra of most octahedral complexes of ions with the configurations $d^1$, $d^4$, $d^6$ and $d^9$ are characterized by a single absorption, while those for many corresponding $d^2$, $d^3$, $d^7$ and $d^8$ complexes have up to three main absorptions.

e. The spectra of $d^5$ complexes show a large number of very weak absorptions, some of which are relatively sharp.

f. The perceived colours of these complexes by transmission are those complementary to the absorptions. Suppose white light impinges upon a sample from a direction labelled $z$. At appropriate absorption frequencies, the electronic

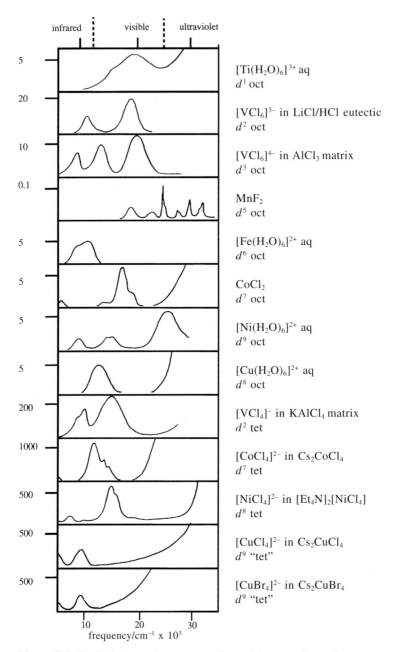

**Figure 2-1.** Typical absorption spectra of transition-metal complexes.

arrangements in the molecules change as energy is absorbed. About $10^{-18}$ seconds later, the same energy (frequency) is re-emitted and the ground state electronic arrangement is recovered. However, the light is emitted equally in all directions normal to the incident electric displacement. When viewed along direction $z$, less of the absorbed frequencies are observed than if no resonance had occurred and we record a net absorption in our spectrum. Further, the colour that we *observe* with our eyes is, of course, determined by absorptions occurring only in the *visible* part of the electromagnetic spectrum.

g.  With the latter point in mind, we note the colours of the permanganate ion (deep purple) and of the tetrachloro- and tetrabromocuprate(II) ions in $Cs_2[CuCl_4]$ and $Cs_2[CuBr_4]$ (yellow and brown). That for the tetrabromo complex is rather intense because the origin of the charge-transfer band lies lower in energy than that for the tetrachloro complex and we could describe the brown colour of $Cs_2[CuBr_4]$ as a sort of 'red-black'.

---

**Box 2-1**

Spectrometers frequently record spectra on a wavelength scale (nm). This is because dispersion by gratings and prisms is more nearly linear in wavelength than in frequency. On the other hand, frequencies of transitions are directly related to the energy changes which are of more chemical significance. We shall report transition energies throughout this book on frequency scales. Frequency and wavelength are reciprocally related and 10,000 wavenumbers ($cm^{-1}$) = 1000 nm.

---

There are many more details to be recognized within even the spectra illustrated in Fig. 2-1: sometimes, we observe bands which have split into two or more components, so that some of the generalizations above are spoilt. We shall look into these matters in some detail in due course. For the moment, there are two main features of all '$d-d$' spectra upon which we must focus:

1) '$d-d$' bands are relatively weak, and
2) the number and patterns of '$d-d$' absorptions are characterized by the molecular geometry and by the $d^n$ configuration.

We shall return to 1) in Chapter 4. Here we consider the significance of the $d^n$ configuration.

## 2.2  The Valence Shell

The species discussed so far belong to the class we might label Werner-type complexes. We use this description to differentiate from carbonyl-type or other low oxidation state complexes. We stay with Werner-type complexes exclusively until Chapter 6. The radial waveforms for $3d$, $4s$ and $4p$ orbitals of the metals in such

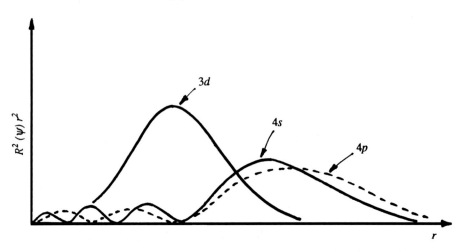

**Figure 2-2.** Schematic representation of the radial waveforms for 3*d*, 4*s* and 4*p* orbitals in first row transition-metal ions of intermediate oxidation state (Werner-type complexes).

complexes are shown qualitatively in Fig. 2-2 and emphasize a most important point. This is that the 3*d* orbitals in *Werner-type complexes* are much more 'inner' than either the 4*s* or 4*p* orbitals. Though hardly core-like, the radial extension of the 3*d* orbitals is not great. Overlap of the metal 3*d* orbitals with ligand functions is correspondingly small. Before claiming that the extent of admixture of the *d* orbitals into the bonding molecular orbitals of a complex is also small, however, we must consider the relative energies of all orbitals involved. We expect the orbital energy ordering for metals in higher oxidation states to be: ligand donor function < metal 3*d* < metal 4*s* < metal 4*p*. On these grounds alone, metal orbital participation in any bonding molecular orbitals formed would be expected to decrease in the order 3*d*> 4*s*> 4*p* as the energy separation between metal and ligand orbitals increases. But as we have noted from Fig. 2-2, the ordering on overlap grounds would be 4*p* ≥ 4*s* > 3*d*.

   These trends are sketched in Fig. 2-3. We argue that while the energy matching favours strongest participation of the 3*d* function amongst the metal functions, these orbitals are sufficiently withdrawn or contracted that their poor overlap with ligand functions leaves the metal 4*s* orbital as the dominant metal contribution in the bonding.

   This idea is a central thesis in this book. We shall re-emphasize the point again and again, and justify our position increasingly as we progress. Here, we make just one or two remarks about it. Firstly, we are not saying that the metal *d* orbitals aren't involved in the orbitals that bind a complex, but merely that their participation is small. It is perfectly possible, however, to develop a consistent picture of chemical bonding, spectral and magnetic properties *together* using, *as a starting point,* the idea that the *d* orbitals have effectively *no* participation in the bonding orbitals. This will seem a strange idea to many since the implication of all teaching texts, so far as we are aware, is that "transition-metal chemistry is about the consequences of

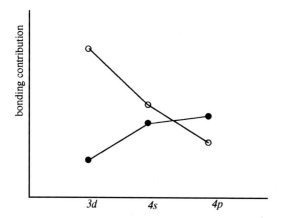

**Figure 2-3.** Contribution to bonding from energy matching with ligand orbitals (○), and from overlap with ligand orbitals (●).

*d*-orbital overlap." We consider that such a view sets up a false prejudice in the mind of the reader and has engendered serious misunderstanding of the subject we call 'ligand-field theory', as we shall discuss. It is to be acknowledged that our assertion that the *d*-orbital participation in the bonding orbitals of a complex is small leaves open the question of 'how small is small?'. As we shall see, however, even the limiting assumption of negligible participation of the *d* orbitals provides a most valuable viewpoint. So, with the promise to return to this seminal question and to refine our position, let us now see something of what follows from the proposition.

The proposition is that the bonds holding a Werner-type complex together are dominated by overlap of unspecified ligand orbitals with the transition-metal *4s* orbital. The latter is, of course, spherically symmetric, so that the attractive (bonding) forces are largely undirected. This bequeaths to secondary repulsive forces, like ligand–ligand repulsions, the determination of the molecular angular geometry. Straightaway, therefore, the reason for the phenomenal success of Kepert's model, as described in the first chapter, is apparent. To be utterly simplistic about it: at this level, the metal doesn't care about the angular geometry, but the ligands do. The picture is very rough, of course, and still fails to explain the existence of planar complexes. We return to *that* question in Chapter 7.

Two other, closely related, consequences flow from our central proposition. If the *d* orbitals are little mixed into the bonding orbitals, then, by the same token, the bond orbitals are little mixed into the *d*. The *d* electrons are to be seen as being housed in an essentially discrete – we say 'uncoupled' – subset of *d* orbitals. We shall see in Chapter 4 how this correlates directly with the weakness of the spectral '*d–d*' bands. It also follows that, regardless of coordination number or geometry, the separation of the *d* electrons implies that the $d^n$ configuration is a significant property of Werner-type complexes. Contrast this emphasis on the $d^n$ configuration in transition-metal chemistry to the usual position adopted in, say, carbon chemistry where $sp$, $sp^2$ and $sp^3$ hybrids form more useful bases. Put another way, while the *2s* and $2p$ subshells *together* comprise the valence shell in carbon chemistry, the $d$ subshell of Werner-type complexes retains a free-ion-like integrity alongside a metal valence shell of $4s$ (with some $4p$) character.


## 2.3  The Roles of $d$ Electrons


Surely a natural question to ask at this stage is 'if the $d$ orbitals essentially don't overlap with the ligand orbitals, what role, if any, do they play?'. Although there is an implication in that question that any role is minor, that is not the case at all. The $d$ electrons interact with the bonding electrons. Let us emphasize the word 'interact': it refers to a *mutual* action. The $d$ electrons are affected by the bonding electrons and the bonding electrons are affected by the $d$ electrons. We can progress a long way by considering these two aspects separately. Ultimately, to be sure, we must refine our arguments to make due recognition of the *inter*action.

The effects of the bonding electrons upon the $d$ electrons is addressed within the subjects we call crystal-field theory (CFT) or ligand-field theory (LFT). They are concerned with the $d$-electron properties that we observe in spectral and magnetic measurements. This subject will keep us busy for some while. We shall return to the effects of the $d$ electrons on bonding much later, in Chapter 7.


## Suggestions for further reading

1. The Roles of $d$-Electrons in Transition Metal Chemistry: A New Emphasis, M . Gerloch, *Coord. Chem. Rev.*, **1990**, *99*, 199.
2. P.W. Atkins, *Molecular Quantum Mechanics,* Oxford University Press, Oxford, **1970**.
3. J.N. Murrell, S.F.A. Kettle, J.M. Tedder, *Valence Theory,* 2nd ed., Wiley, New York, **1969**.
   – The last two references have more to say on the radial forms of atomic orbitals.

# 3  Crystal-Field Splittings

## 3.1  The Crystal-Field Premise

During the first twenty years or so of this century, an incredibly detailed understanding of atomic line spectra was built up with the application of the, then new, quantum theory. Indeed, the development of quantum theory came about in part by the need to understand these spectral properties. We shall have to review some basic features of the theory of atomic spectra for our present purposes, but we shall leave it for the moment.

In the later 1920's, physicists, rightly flushed with their successes with interpreting the rich, sharp spectra of atoms and gas phase ions, sought to extend their reach to the broader (and fewer) absorption bands that characterize the spectra of ions in crystalline matrices.* These bands occur at utterly different frequencies to those of the corresponding free ions so that there is no similarity at all between the spectra of free ions and of those in ionic or covalent lattices.

Crystal-field theory (CFT) was constructed as the first theoretical model to account for these spectral differences. Its central idea is simple in the extreme. In free atoms and ions, all electrons, but for our interests particularly the 'outer' or non-core electrons, are subject to three main energetic constraints: a) they possess kinetic energy, b) they are attracted to the nucleus and c) they repel one another. (We shall put that a little more exactly, and symbolically, later). Within the environment of other ions, as for example within the lattice of a crystal, those electrons are expected to be subject also to one further constraint. Namely, they will be affected by the non-spherical electric field established by the surrounding ions. That electric field was called the 'crystalline field', but we now simply call it the 'crystal field'. Since we are almost exclusively concerned with the spectral and other properties of positively charged transition-metal ions surrounded by anions of the lattice,** the effect of the crystal field is to *repel* the electrons.

Those electrons must not only avoid each other but also the negatively charged anionic environment. In its simplest form, the crystal field is viewed as composed of an array of negative point charges. This simplification is not essential but perfectly adequate for our introduction. We comment upon it later.

---

* It is interesting that the very broad, so-called 'spin-allowed' transitions, like most of those in Fig. 2-1, were not actually recognized as such until the 1950's. This was because of the characteristics of the spectrograph rather than the spectrometer.

** To be contrasted with a negatively charged metal surrounded by positively charged groups. The idea of neutral ligands with donor lone pairs will be considered in due course.

## 3.2  Splitting of *d* Orbitals in Octahedral Symmetry

We are concerned with what happens to the (spectral) *d* electrons of a transition-metal ion surrounded by a group of ligands which, in the crystal-field model, may be represented by point negative charges. The results depend upon the number and spatial arrangements of these charges. For the moment, and because of the very common occurrence of octahedral coordination, we focus exclusively upon an octahedral array of point charges.

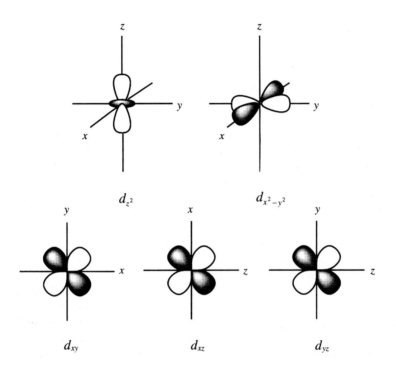

**Figure 3-1.** The angular forms of the five *d* orbitals.

The set of five *d* orbitals share a common radial part like that sketched in Fig. 2-2. Their angular parts are shown in Fig. 3-1. Let us consider the six point charges in an octahedral array to be disposed along the positive and negative *x*, *y* and *z* axes to which these *d* orbitals are referred. This is conveniently drawn, as shown in Fig. 3-2, by placing the charges at the centres of each face of a cube, itself centred on the metal atom. By comparing the orbitals in Fig. 3-1 with the crystal field of point charges in Fig. 3-2, we observe that some orbitals are more directed towards the point charges than others. The $d_{z^2}$ and $d_{x^2-y^2}$ orbitals are directed exactly towards the six charges while the $d_{xy}$, $d_{xz}$ and $d_{yz}$ have lobes which lie between the *x*, *y* and *z*

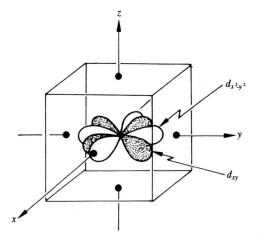

**Figure 3-2.** The $d_{x^2-y^2}$ orbital points towards (four of) the ligands and $d_{xy}$ points between those ligands.

axes on which the charges are situated. Therefore, an electron placed in the $d_{xy}$ orbital will be less repelled by the crystal field than one placed in the $d_{x^2-y^2}$. It is obvious that the dispositions of the $d_{xy}$, $d_{yz}$ and $d_{xz}$ orbitals with respect to the point charges are energetically equivalent. It is not obvious, but nonetheless true, that the repulsion suffered by an electron in the $d_{x^2-y^2}$ orbital is the same as that by one in the $d_{z^2}$ orbital (see Box 3-1).

Altogether then, the energies of the five $d$ orbitals (strictly of the electrons within them) in octahedral symmetry separate into two groups as shown in Fig. 3-3. *All d orbitals are raised in energy by repulsion in the crystal field, but two go to higher energies than the other three.* Since spectroscopy and, indeed, most other $d$ electron properties of interest to us are concerned with relative energies, or splittings, rather than with absolute energies, a more usual representation of the differential crystal-field effect upon $d$ orbitals is that shown in Fig. 3-4. We draw the energy levels with respect to the mean energy of the whole $d$ orbital set. This so-called

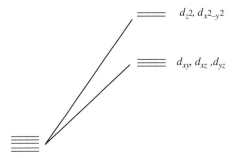

**Figure 3-3.** Two $d$ orbitals are raised in energy more than the other three.

**Box 3-1**

Each lobe of the $d_{x^2-y^2}$ orbital interacts predominantly with one point charge. The repulsive effects relate to the electron *density* within any given orbital so we might describe the interaction in units of 'lobe repulsion' and say that, for the $d_{x^2-y^2}$ orbital, this amounts to $4^2$ = 16 repulsion units (4 *squared* because electron density $\propto \psi^2$).

The $d_{z^2}$ orbital can be written as a linear combination of two different orbitals which (Eq. 3.1) look like the $d_{x^2-y^2}$ orbital but referred to the *xz* and *xy* planes: $d_{z^2-x^2}$ and $d_{z^2-y^2}$

$$d_{z^2} = (d_{z^2-x^2} + d_{z^2-y^2})/\sqrt{2} \qquad (3.1)$$

This identity is sketched below.

The $d_{z^2-x^2}$ and $d_{z^2-y^2}$ orbitals each interact with four point charges in precisely the same way as does the $d_{x^2-y^2}$ orbital. Again the repulsion relates to electron density, so the total interaction of the combination is $(4/\sqrt{2})^2 + (4/\sqrt{2})^2 = 16$ of our repulsion units. In other words, the $d_{z^2}$ and $d_{x^2-y^2}$ orbitals are degenerate in octahedral symmetry.

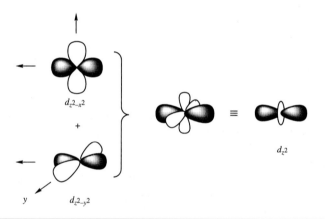

'barycentre' rule means that if the splitting between the two subsets of orbitals is labelled $\Delta_{oct}$, the higher pair lie at an energy $+0.6\,\Delta_{oct}$ and the lower trio at $-0.4\,\Delta_{oct}$. ('Barycentre' means a 'centre-of-gravity' type of rule.) An older alternative label for the octahedral-field splitting is $10Dq$ where, in the literal crystal-field model we have introduced thus far, $q$ is the charge on each ligand and $D$ is a quantity related to the geometry. We shall make no use of these old meanings and just refer to $Dq$ as a sort of 'dipthong of consonants'. In this notation, the pair of orbitals lies at $+6\,Dq$ and the lower trio at $-4\,Dq$. Both $\Delta_{oct}$ and $10Dq$ are in common use and we shall swap between them at will.

The subsets of *d* orbitals in Fig. 3-4 may also be labelled according to their symmetry properties. The $d_{z^2}/d_{x^2-y^2}$ pair are labelled $e_g$ and the $d_{xy}/d_{xz}/d_{yz}$ trio as $t_{2g}$. These are group-theoretical symbols describing how these functions transform under various symmetry operations. For *our* purposes, it is sufficient merely to recognize that the letters *a* or *b* describe orbitally (*i.e.* spatially) singly degenerate species, *e* refers to an orbital doublet and *t* to an orbital triplet. Lower case letters are used for one-electron wavefunctions (*i.e.* orbitals). The *g* subscript refers to the behaviour of

**Box 3-2**

Another way to view the barycentre rule is to consider first the bringing up to the metal of a spherical shell of negative charge which increases the energies of all five $d$ orbitals equally. Then, in this notional picture, if the spherical shell of charge redistributes towards the apices of an octahedron, those orbitals directed towards those apices suffer a further repulsion and energy increase, while those directed in between, acquire a relative stability.

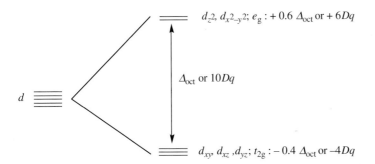

$d_{z^2}, d_{x^2-y^2}; e_g : + 0.6\ \Delta_{oct}\ \text{or} + 6Dq$

$\Delta_{oct}$ or $10Dq$

$d$

$d_{xy}, d_{xz}, d_{yz}; t_{2g} : -0.4\ \Delta_{oct}\ \text{or} -4Dq$

**Figure 3-4.** Barycentre splitting of the $d$ orbitals in octahedral symmetry.

these functions under inversion through a centre – *gerade* or *even*. As all $d$ orbitals are centrosymmetric, all $d$ subsets are here labelled $g$. The subscript 2 in $t_{2g}$ gives further symmetry information. We do not require this here and must just accept the label as a name.

For a transition-metal ion in an octahedral environment with a single $d$ electron (configuration $d^1$), the ground state arises when that electron is placed in the lower energy $t_{2g}$ subset. Upon absorption of an appropriate energy – $\Delta_{oct}$, the electron is promoted into the higher energy $e_g$ subset. Redistribution of the electron within the $t_{2g}$ set involves no energy change and will take place spontaneously and continuously because of the equivalence of the three cartesian directions in the octahedron. Thus, only one energy change within the $d$ orbitals is possible and corresponds to the *transition* $t_{2g} \rightarrow e_g$. Illuminating such $d^1$ ions with light of varying frequency, as in spectroscopy, may bring about that single transition when the frequency $h\nu$ is such that $h\nu = \Delta_{oct}$. The '$d-d$' spectrum of the $d^1$ ions comprises this single optical transition.

Yet another representation of the $d$ orbital splitting in Fig. 3-4 is that shown in Fig. 3-5. Here, we imagine that the charges on the six ligands are smoothly varied or that their distances from the metal atom are so varied or that the radial extension of the $d$ orbitals is changed; or, of course, any combination of these. The splitting *pattern* ($t_{2g} + e_g$) remains unchanged for it is a property of the octahedral disposition of the point changes. The *magnitude* of the splitting parameter $\Delta_{oct}$, however changes as the strength of the repulsions between $d$ electrons and point charges changes. As such $\Delta_{oct}$ thus measures the *strength* of the crystal field.

Now suppose the system of metal and charges we have discussed represents a metal ion complexed by six ligands. These vibrate continuously. One such vibration

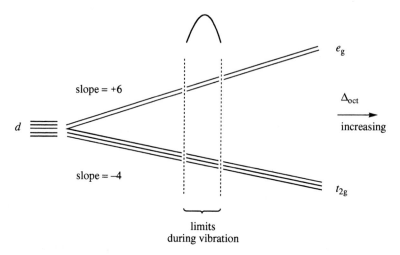

**Figure 3-5.** The variation of the octahedral splitting with respect to smoothly changing the magnitude of $\Delta_{\text{oct}}$.

involves the bonds lengthening and shortening together – a so-called 'breathing' mode. When the ligands are closer to the metal, the repulsions suffered by the *d* electrons are larger than when the ligands are more distant. So, during the course of the vibration, the crystal field strength varies between the limits indicated in Fig. 3-5. Such vibrations typically take place in about $10^{-13}$ sec. An optical event involving absorption and reemission between the $t_{2g}$ and $e_g$ subsets takes place in about $10^{-18}$ sec. A beam of light, incident on a sample, therefore effectively 'sees' stationary molecules (the Frank-Condon principle). In a sample containing many such molecules, however, the light encounters molecules in every part of their vibrational cycles and so we observe electronic '*d–d*' transitions at frequencies everywhere between the extreme vibration limits indicated in Fig. 3-5. Simple theory predicts that $\Delta_{\text{oct}}$ is inversely proportional to the fifth power of the bond length (more sophisticated calculations actually give rather similar results). The energy spread indicated in Fig. 3-5 is therefore actually rather large and '*d–d*' transitions are typically observed as broad bands, maybe 2000 to 3000 cm$^{-1}$ wide.

## 3.3 Splitting of *d* Orbitals in Tetrahedral and Other Symmetries

*Tetrahedral Symmetry*
Perhaps only slightly less common than octahedral symmetry is tetrahedral symmetry. We now examine the *d* orbital splitting in this environment. The story is much the same as above, except that it is now convenient to place the four point charges of the tetrahedron as shown in Fig. 3-6. Here ligands are put at alternate

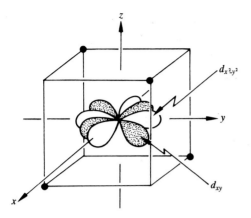

**Figure 3-6.** The angle subtended by a lobe of $d_{x^2-y^2}$ and M – L is greater than that subtended by a lobe of $d_{xy}$ and M – L.

corners of a cube centred about the metal. We retain the same axis frame so that $x$, $y$ and $z$ are again directed towards the midpoints of the cube faces. The $d_{xy}$ and $d_{x^2-y^2}$ orbitals are included in the figure for discussion (the shading has no significance other than to differentiate these two orbitals).

We note that the lobes of $d_{xy}$ are directed towards the midpoints of the cube edges, whereas those of $d_{x^2-y^2}$ point towards the midpoints of the cube faces. It is apparent that the lobes of $d_{xy}$ are oriented more nearly towards the point charges than are those of $d_{x^2-y^2}$. An electron in $d_{xy}$ is thus repelled more than one in $d_{x^2-y^2}$. Once again, it is obvious that the situations for $d_{xz}$ and $d_{yz}$ electrons are entirely like that for $d_{xy}$ (permute the axis labels again) so these orbitals form a subset of three. Less obviously, the $d_{z^2}$ and $d_{x^2-y^2}$ are equivalent and form a subset of two. The splitting diagram for the five $d$ orbitals in a tetrahedral crystal field is shown in Fig. 3-7. Once more, orbital energies are indicated with respect to a barycentre rule. The splitting is called $\Delta_{tet}$ (or $10Dq$ – which can be confusing; see Section 3.11) and the trio lies higher in energy than the pair. The $d$ orbital subsets are labelled $t_2$ and $e$.

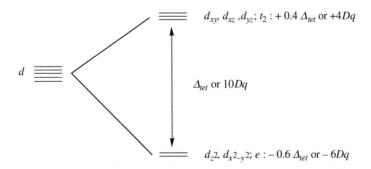

**Figure 3-7.** Barycentre splitting of $d$ orbitals in tetrahedral symmetry.

Note the absence of the g subscripts here. Although the d orbitals are still centro-symmetric, the tetrahedral environment lacks a centre of inversion. The d orbitals are therefore not classified with respect to a symmetry element which doesn't exist: the absence of the g subscript does *not* imply the opposite – i.e. u *(ungerade or odd)*.

Overall, then, we observe that the orbitals are inversely split in the tetrahedron with respect to the octahedron. However, the differentiation between the subsets in the octahedron was based on orbitals being oriented directly at or between the point charges; in the tetrahedron, all d orbitals point between the ligands, though some are closer to the point charges than others. Accordingly, the magnitude of the tetrahedral splitting is less than that in the octahedron. Simple geometrical calculations show that, for the *same* metal (same d orbital radial functions) *and* for the *same* bond lengths, these splittings are related by the expression in Eq. (3.2).*

$$\Delta_{tet} = 4/9\ \Delta_{oct} \tag{3.2}$$

In practice, these conditions of radial waveforms and bond lengths will not be met exactly, so that a rough rule is that $\Delta_{tet} \approx 0.5\ \Delta_{oct}$ in real systems. Once again, only one electronic, '$d-d$', absorption is expected (and observed), although much shifted towards the red relative to that in an analogous octahedral complex.

*Other Environments*
The splitting patterns in crystal fields of symmetries other than octahedral or tetrahedral can be worked out using broadly similar principles. In general, the d-orbital degeneracy is raised (meaning decreased!) even more and there result up to five energetically discrete subsets. That in turn begets more splitting parameters like $\Delta_{oct}$ and often the situation ceases to be simple (see Box 3-3). We shall look at some of these situations briefly in later chapters. For the moment, it suffices to restrict our concern to the so-called 'cubic symmetries' – octahedral and tetrahedral. So many complexes possess these symmetries, or at least approximately, that this restriction is not too serious at this stage.

## 3.4  Holes: $d^1$ and $d^9$

The discussion above might have pertained for example, to the energies and electronic spectra of titanium(III) compounds. The same ideas can be applied with just one modification to the d-electron properties of copper(II) complexes and other

---

* Some prefer to write $\Delta_{tet} = -\ ^4/_9\ \Delta_{oct}$ in order to emphasize the inversion of t and e orbital subsets. However, if $\Delta_{oct}$ and $\Delta_{tet}$ are defined as the orbital *splittings*, it is probably best to omit the sign.

**Box 3-3**

As an example of the effects of lower symmetry, consider the splitting of the *d*-orbital energies in a tetragonally elongated octahedron. This could arise either through the obvious arrangement of two long *trans* contacts (bond lengths) and four short, or by ligand dissimilarities as in a *trans* $MA_4B_2$ complex.

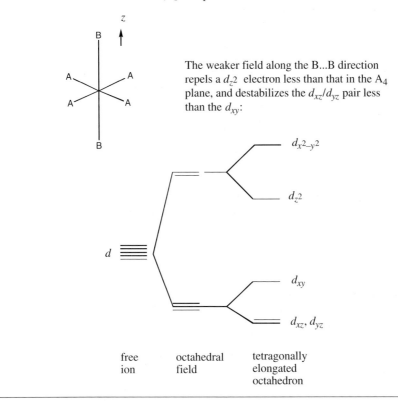

The weaker field along the B...B direction repels a $d_{z^2}$ electron less than that in the $A_4$ plane, and destabilizes the $d_{xz}/d_{yz}$ pair less than the $d_{xy}$:

free ion     octahedral field     tetragonally elongated octahedron

$d^9$ ions. Consider an octahedral $d^9$ complex with the so-called 'strong-field' configuration $t_{2g}^6 e_g^3$ as shown on the left side of Fig. 3-8. This electronic arrangement, or configuration, clearly corresponds to the electronic ground state for the lower-lying $t_{2g}$ orbital set being filled while the higher-lying $e_g$ set is incompletely filled.

On absorption of an energy $\Delta_{oct}$, one of the $t_{2g}$ electrons will be promoted into the $e_g$ set, as on the right side of Fig. 3-8. As the $e_g$ set is now full, no further electronic promotions are possible so that this corresponds to the one and only excited state of the octahedral $d^9$ configuration. We thus observe a single absorption band in the '$d-d$' spectrum. The excitation $t_{2g}^6 e_g^3 \rightarrow t_{2g}^5 e_g^4$ is equivalent to the transfer of the hole in the $e_g^3$ configuration into the $t_{2g}$ set. We may view a transition in a $d^9$ complex as a redistribution of a single hole (lack of electron) within a full $d$ shell. However, while the transition of a single electron in octahedral $d^1$ complexes involves the shift $t_{2g} \rightarrow e_g$, the transition of a single hole in the corresponding $d^9$ system involves the shift $e_g \rightarrow t_{2g}$. A similar inversion follows for tetrahedral complexes. We shall

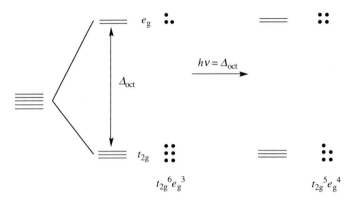

**Figure 3-8.** The electronic transition for octahedral $d^9$ ions.

have more to say about this hole formalism later. For the moment, it suffices to note that octahedral and tetrahedral complexes of $d^1$ or $d^9$ configurations give rise to just *one* electronic transition in their '$d-d$' spectra. Some of the general features of Fig. 2-1 are thus explained. Unfortunately, the explanation of the greater number of transitions in, say, $d^2$ complexes is more difficult and lengthy. We must look at that, however: it is our next topic. In mastering it we shall enter quite deeply into the subject of crystal-field theory and learn more about the energetics of $d$ electrons in transition metal complexes.

## 3.5  More Transitions for $d^2$

Let us look now at the case of a $d^2$ ion in octahedral symmetry. The orbital splitting is again as given in Fig. 3-4. With two $d$ electrons, however, rather more electronic arrangements within the $t_{2g}$ and $e_g$ subsets are possible. At first sight there are three: both electrons may be housed in the $t_{2g}$ subset, both in the $e_g$ or one electron in each, as indicated on the left of Fig. 3-9. In fact there are many more and they arise for two reasons. The first concerns spin, since the electrons could either share a common spin or their spins could be opposed. In the former case, the total spin quantum number for the pair of electrons is $(\frac{1}{2} + \frac{1}{2})$ or 1 while, in the latter, it is $(\frac{1}{2} - \frac{1}{2})$ or 0. The two-electron states associated with these spin quantum numbers are called (spin) triplets or singlets respectively. For those somewhat unfamiliar with these labels, we provide a brief review in the next section. Now, in this section, we consider only those electronic arrangements of maximum spin. We do this for two reasons. Firstly, Hund's first rule defines the ground term of a free ion to be one of maximum spin multiplicity (see Section 3.6). Secondly, as will be discussed in Chapter 4, electronic transitions between states of the same spin multiplicity are much more allowed (the spectral bands more intense) than those involving a change of spin. So we focus here on the spin-triplet states. However, even discarding the

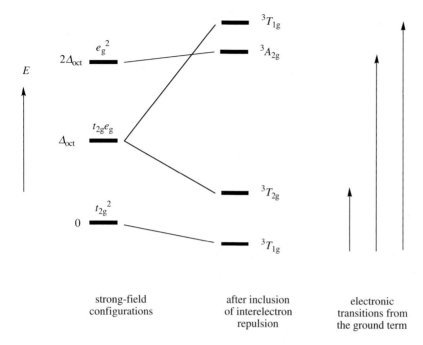

$E$

strong-field configurations

after inclusion of interelectron repulsion

electronic transitions from the ground term

**Figure 3-9.** Four spin-triplet terms arise for $d^2$.

spin-singlet states, there arise four discrete types of spin-triplet arrangement rather than the three one might at first anticipate. Let us see why.

Firstly, consider the spatial degeneracies of spin-parallel electronic arrangements within the configurations $t_{2g}^2$, $t_{2g}^1 e_g^1$, $e_g^2$. Parallel spins must be placed in different orbitals, of course, because of the Pauli exclusion principle. So spin-parallel arrangements within the $t_{2g}^2$ configuration necessarily involve one up-spin (say) electron in each of two of the members of the $t_{2g}$ orbital set. There are three such arrangements. (The same result is to be had by noting that the empty orbital can be any one of three). We label this group of arrangements by the crystal-field term symbol, $^3T_{1g}$ (see Box 3-4).

Next we consider the configuration $e_g^2$. A spin-parallel arrangement must involve one electron in each of the two members of the $e_g$ subset. Ignoring spin (for we

---

**Box 3-4**

The left superscript indicates that the arrangements are all spin triplets. The letter $T$ refers to the three-fold degeneracy just discussed and it is in upper case because the symbol pertains to a many-electron (here two) wavefunction (we use lower-case letters for one-electron wavefunctions or orbitals, remember). The subscript $g$ means the wavefunctions are even under inversion through the centre of symmetry possessed by the octahedron (since each $d$ orbital is of $g$ symmetry, so also is any product of them), and the right subscript 1 describes other symmetry properties we need not discuss here. More will be said about such term symbols in the next two sections.

---

**Box 3-5**

Again the left superscript indicates the spin-triplet nature of the arrangement. The letter $A$ means that it is spatially (orbitally) one-fold degenerate and it is upper-case because we describe two-electron wavefunctions. The subscript is $g$ because the product of $d$ orbitals is even under the octahedral centre of inversion, and the right subscript 2 must remain a mystery for us once again.

---

have fixed that as spin-parallel), there is only one way of filling this orbital subset. This unique arrangement is labelled by the term symbol $^3A_{2g}$ (see Box 3-5).

As the energy of the $e_g$ orbital set in the octahedron is larger than that of the $t_{2g}$, the energy of the configuration $e_g^2$ is greater (by $2\Delta_{oct}$, that is, $\Delta_{oct}$ for each electron) than that of $t_{2g}^2$. It should not be surprising that the energy of the $^3A_{2g}$ term (group of wavefunctions) is higher than that of the $^3T_{1g}$ term discussed above. Note, however, that the energy of the $^3A_{2g}$ term is not determined solely by the $2\Delta_{oct}$ promotion as we shall see.

We have left the configuration $t_{2g}^1 e_g^1$ till last because it involves some new ideas. Actually, as will be apparent, all three configurations involve exactly the same principles. While their variety is not immediately apparent from our discussion of the $e_g^2$ and $t_{2g}^2$ configurations, it is with the $t_{2g}^1 e_g^1$. Now the $t_{2g}^1 e_g^1$ configuration symbol means that one electron is to be placed within the $t_{2g}$ subset and one within the $e_g$. We have agreed to consider only those arrangements with parallel spins. The $t_{2g}$ electron may be housed in any one of three orbitals while *independently* the $e_g$ electron may occupy one of two. Altogether therefore, there are (3 x 2) = 6 spatial arrangements for these two electrons. However, the six arrangements are not degenerate. They form up into two sets of three, with the term symbols $^3T_{2g} + {}^3T_{1g}$ (the rules of the labelling are now hopefully clear enough). Why are the energies of these two terms different? Also of note is that the difference in their energies is not at all trivial, being around twice the magnitude of the crystal-field splitting $\Delta_{oct}$ in many systems.

The answer has been given in Section 3.1. We have focused upon *crystal-field* energies, that is, upon the need of the metal $d$ electrons to avoid the regions of higher negative charge in the crystal (or molecular) environment. With more than one $d$ electron (or hole), as here, we must not forget that these electrons also need to avoid each other. Our discussion in the present section thus far has omitted consideration of *interelectron repulsion* energy. Recall our caveat when asserting that the relative energy of the $^3T_{1g}$ (from $t_{2g}^2$) and the $^3A_{2g}$ (from $e_g^2$) is not given simply by $\Delta_{oct}$, the energy splitting of the $t_{2g}$ and $e_g$ orbitals. That is because the electron–electron repulsion energies for these two arrangements are not the same. They are not the same because the spatial proximity of members of the $e_g$ orbital pair is not the same as that for members of the $t_{2g}$ set. Similarly, the proximities and interelectron repulsion energies vary within the electronic arrangements of the $t_{2g}^1 e_g^1$ configuration, as we now discuss.

Taking one electron from each of the $t_{2g}$ and $e_g$ subsets, we can form high- and low-energy spatial triplets, $^3T_{2g}$ and $^3T_{1g}$, as shown in Eq. (3.3) (we write $xy$ for $d_{xy}$, etc).

$${}^3T_{2g}: \quad (xy)(z^2) \qquad (yz)(x^2-y^2) \quad (xz)(x^2-y^2)$$

$$(3.3)$$

$${}^3T_{1g}: \quad (xy)(x^2-y^2) \quad (yz)(z^2) \qquad (xz)(z^2)$$

The ${}^3T_{2g}$ term wavefunctions lie lower in energy than those of the ${}^3T_{1g}$. Qualitatively, we can rationalize this energy ordering most easily by considering the first component of each of these terms. The relevant pairs of orbitals are shown in Fig. 3-10. The orbital pair $(xy)(x^2-y^2)$ is obviously much more crowded together than is the $(xy)(z^2)$ pair and so the interelectron repulsion energy associated with the former arrangement is much greater than with the latter.

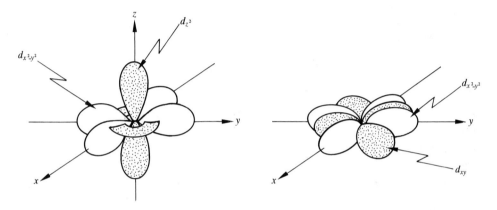

**Figure 3-10.** Relative electron crowding for different orbital pair densities.

On the right side of Fig. 3-9 are represented the relative energies of the two ${}^3T_{1g}$ terms, the ${}^3T_{2g}$ and ${}^3A_{2g}$. The ground term is the ${}^3T_{1g}$ from the $t_{2g}^2$ configuration. Spin-allowed electronic transitions (those between terms of the same spin angular momentum – but see also Sections 3.6, 3.7 and Chapter 4) now take place upon excitation from ${}^3T_{1g} \rightarrow {}^3T_{2g}$, $\rightarrow {}^3A_{2g}$, $\rightarrow {}^3T_{1g}$. The '$d$–$d$' spectra of octahedrally coordinated $d^2$ ions thus exhibit *three* bands. Similar arguments for tetrahedrally coordinated $d^2$ ions yield three transitions also, but this time from a ${}^3A_2$ ground term: ${}^3A_2 \rightarrow {}^3T_2$, $\rightarrow {}^3T_1$, $\rightarrow {}^3T_1$. Clearly, similar results apply to those $d^8$ ions having two holes in a full $d$ shell rather than two electrons. We shall look at these hole equivalencies more carefully in Section 3.7.

The approach we have adopted for the $d^2$ configuration began from the so-called 'strong-field' limit. This is to be contrasted to the 'weak-field' scheme that we describe in Section 3.7. In the strong-field approach, we consider the crystal-field splitting of the $d$ orbitals first, and then recognize the effects of interelectron repulsion. The opposite order is adopted in the weak-field scheme. Before studying this alternative approach, however, we must review a little of the theory of free-ion spectroscopy

## 3.6　Atomic Orbitals and Terms

The present section is offered as a *review* of the jargon of the theory of free-ion spectroscopy with little in the way of *any free-standing explanation.*[*]
　　A transition metal with the configuration $d^1$ is an example of a 'hydrogen-like' atom in that we consider the behaviour of a single ($d$) electron outside of any closed shells. This electron possesses kinetic energy and is attracted to the shielded nucleus. The appropriate energy operator (Hamiltonian) for this is shown in Eq. (3.4).

$$\mathcal{H}_{H-like} = -\frac{\hbar^2}{2m}\nabla^2 - \frac{e^2}{r} \tag{3.4}$$

Solutions to the Schrödinger equation (3.5) are called *one-electron* wavefunctions or *orbitals* and take the form in Eq. (3.6)

$$\mathcal{H}_{H-like}\phi = \varepsilon\,\phi \tag{3.5}$$

$$\phi = R_{nl}(r)Y_m^l(\theta,\varphi) \tag{3.6}$$

The radial functions, $R$ depend only upon the distance, $r$, of the electron from the nucleus while the angular functions, $Y_m^l(\theta,\varphi)$, called *spherical harmonics*, depend only upon the polar coordinates, $\theta$ and $\varphi$. Examples of these purely angular functions are shown in Fig. 3-11.

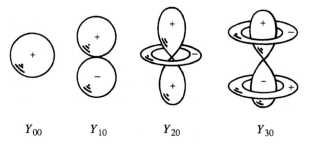

$Y_{00}$　　　　$Y_{10}$　　　　$Y_{20}$　　　　$Y_{30}$

**Figure 3-11.** The shapes of the angular functions are determined only by the theory of angular momentum in spherical symmetry.

　　The orbitals are labelled 1s; 2s, 2p; 3s, 3p, 3d; etc. to indicate the principal quantum number $n$ (here equal to 1, 2, 3.....) and the orbital angular momentum quantum number, $l$, according to the code in (3.7).

---

[*] This topic is described fully, but at the same level as adopted in the present book, in 'Orbitals, Terms and States'.

$$l = \quad 0 \quad 1 \quad 2 \quad 3 \quad 4 \quad 5 \quad \ldots \qquad (3.7)$$
$$\phantom{l = \quad} s \quad p \quad d \quad f \quad g \quad h \quad \ldots$$

Associated with each $l$ value are $(2l + 1)$ values of $m_l$, ranging $l$, $l-1$......$-l$, describing the $z$ component of the angular momentum. Thus, we find one $s$ function, three $p$ functions, five $d$ functions, and so on.

The Hamiltonian (3.4) is a function of the usual spatial coordinates ($x$, $y$, $z$ or $r$, $\theta$, $\phi$). Electrons possess the intrinsic property of spin, however, which is to be thought of as a property in an independent, or orthogonal, space (spin space). Spin is actually a consequence of the theory of relativity but we shall merely graft on the property in an *ad hoc* fashion. The spin, $s$, of an electron (don't confuse with $s$ orbitals!) takes the value 1/2 only. The $z$ component of spin, $m_s$, takes $(2s + 1)$ values of $m_s$, ranging $s$, $s-1$,...$-s$. Thus for the single electron, $m_s = +1/2$ or $-1/2$, also labelled $\alpha$ or $\beta$, or indicated by $\uparrow$ or $\downarrow$.

Now consider a $d^2$ ion as an example of a so-called 'many-electron' atom. Here, each electron possesses kinetic energy, is attracted to the (shielded) nucleus *and* is repelled by the other electron. We write the Hamiltonian operator for this as follows:

$$\mathcal{H}_{many} = \sum_i^n \mathcal{H}_{H\text{-}like}(i) + \sum_{i<j}^n \frac{e^2}{r_{ij}} \qquad (3.8)$$

where each pair of the $n$ electrons in the $d^n$ configuration suffer mutual repulsions that are inversely proportional to the instantaneous distance, $r_{ij}$, between them. The sum in the second part is for $i < j$ in order not to count these pair-wise interactions twice and to prevent the $i^{th}$ electron repelling itself. This second operator is called the *Coulomb* operator. Solutions, $\Psi$ for this Hamiltonian,

$$\mathcal{H}_{many} \Psi = \mathcal{E} \Psi \qquad (3.9)$$

are called 'many-electron' wavefunctions, or *term wavefunctions*, because they describe the behaviour of many ($n$) electrons as a group. They are *not* orbitals. These groups of wavefunctions – *terms* – possess the qualities of orbital- and spin-angular momentum, just like the orbitals of (3.5), however. Their orbital angular momentum is labelled by $L$, according to the code:

$$L = \quad 0 \quad 1 \quad 2 \quad 3 \quad 4 \quad 5 \quad \ldots \qquad (3.10)$$
$$\phantom{L = \quad} S \quad P \quad D \quad F \quad G \quad H \quad \ldots$$

and associated with each $L$ value are $(2L + 1)$ values of $M_L$, referring to the $z$ components of the orbital angular momentum.

The spin angular momentum of a term is labelled with $S$ and may take integrally separated values based on 0 or 1/2 depending upon the $d^n$ configuration; *viz.* $S = 0$, 1, 2... or $S = 1/2, 3/2, 5/2$... Associated with each such $S$ value are $(2S + 1)$ values of $M_S$ for the $z$ components of spin angular momentum, with $M_S$ taking the values $S$, $S-1$...$-S$.

The total form of the many-electron wavefunctions, $\psi$, of (3.9) *can* be computed for free ions but only after lengthy numerical procedures. Let us imagine this to have been done and ask what angular momentum properties are assocated with these solutions. We find them aggregating into groups – terms – characterized by appropriate pairs of orbital- and spin-angular momenta: $L$, $S$. Instead of so labelling them, they are conventionally described by *term symbols* of the form $^{2S+1}L$. Examples are $^2D, ^3F, ^6S$ etc, pronounced doublet $D$, triplet $F$, sextet $S$, etc. Each of these terms is $(2L + 1)(2S + 1)$-fold degenerate because there are $(2L + 1)$ $M_L$ values for each $L$ and $(2S + 1)$ $M_S$ values for each $S$; and spin- and orbital-angular momenta are independent properties (in the absence of spin-orbit coupling). The degeneracies of $^2D$, $^3F$, $^6S$ terms, for example, are 10, 21, 6.

Note that, throughout this discussion, we have used lower-case letters when refering to orbitals and upper-case when we mean many-electron wavefunctions. There arises the question of, 'what are the relationships between $l$ and $L$, or between $s$ and $S$ ?'. They are determined by the *vector coupling rule*. This states that the angular momentum for a coupled (*i.e.* interacting) pair of electrons may take values ranging from their sum to their difference (Eq. 3.11).

$$L \rightarrow l_1 + l_2, \quad l_1 + l_2 - 1, \ldots . |l_1 - l_2| \tag{3.11}$$

The same[*] goes for spin angular momentum (Eq. 3.12).

$$S \rightarrow s_1 + s_2, s_1 + s_2 - 1, \ldots .|s_1 - s_2| \tag{3.12}$$

The rule means that the angular momentum of each of a pair of electrons may be parallel – meaning about the same physical axis and in the same sense – or opposed, and that the quantum condition allows only integrally separated values between these limits. For the $d^2$ case, the $l$ value for each electron is 2 and so $L$ can take the values 4, 3, 2, 1 or 0, corresponding to the labels $G, F, D, P, S$. The $s$ value for each electron is 1/2, so $S$ can take values 1 or 0 and there may thus arise terms $^3G, ^3F,$ $^3D, ^3P, ^3S; ^1G, ^1F, ^1D, ^1P, ^1S$. However, not all these terms do arise for the 'equivalent' $d$ electrons of the $d^2$ configuration, because some electronic arrangements violate Pauli's exclusion principle while others ignore the fact that electrons are indistinguishable. For the $d^2$ configuration, the first electron may have any one of five $m_l$ values and any one of two $m_s$ – or 10 possibilities altogether. The second electron is left with only 9 choices of the $(m_l, m_s)$ combination because of the exclusion principle – hence yielding 90 arrangements for the pair. However, the indistinguishability of electrons means we cannot assign meaning to the adjectives 'first' and 'second' here and have thus counted each arrangement twice. Altogether, therefore, the $d^2$ configuration is 45-fold degenerate. It can be shown to give rise to the term set: $^1G, ^3F, ^1D, ^3P, ^1S$. The $(2L + 1)(2S + 1)$-fold degeneracies of these terms are 9, 21, 5, 9 and 1 respectively, adding to 45, as required.

---

[*] The vector coupling rule applies to all forms of angular momentum:
$j \rightarrow j_1 + j_2, j_1 + j_2 - 1, \ldots .|j_1 - j_2|$
  where each $j$ can be $s, S, l, L, j$ or $J$.

configuration                    terms

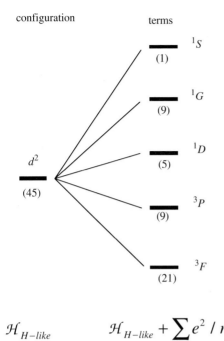

**Figure 3-12.** Free ion terms arising from the $d^2$ configuration.


All this is summarized in Fig. 3-12. The energy ordering of the free-ion terms is *not* determined by consideration of angular momentum properties alone and in general yields only to detailed numerical computation. The ground term – and *only* the ground term – may be deduced, however, from some simple rules due to Hund.

Hund's first rule: The ground term will be one of maximum spin-multiplicity (maximum $S$)

Hund's second rule: If ambiguity remains, the maximum-spin ground term will then be one with maximum $L$

So for $d^2$, the ground term is a spin triplet, and $^3F$ rather than $^3P$. Let us recap one or two points. A configuration describes an orbital assembly in which no recognition is made of the Coulomb interaction between electrons: there are 45 equally good (equi-energetic) ways of arranging two $d$ electrons with regard to their kinetic energy and attraction to the nucleus. When we recognize that these electrons repel and otherwise interact through the Coulomb operator, we observe the term splitting on the right-side of Fig. 3-12. There are now 21 ways of equal minimum energy in which two $d$ electrons possess kinetic energy, are attracted to the nucleus, *and* avoid each other (as best they might). For the (assumed) ordering in Fig. 3-12, we note that there are 9 second-best ways of satisfying these three constraints, and so on. The 21 best ways happen, as it were, to share the common angular momentum properties, designated by the term symbol $^3F$, and so on again. Each member (state) of each term is a two-electron wavefunction, describing one or another particular arrangement of exactly two electrons.

The electrons have coupled through the agency of electrostatic (repulsive) interactions. Electrons may also couple via the (independent) means of magnetic interaction we call *spin-orbit coupling*. This effect is generally smaller than the electrostatic coupling and we shall largely ignore it in this book. Spin-orbit coupling is not unimportant, especially when one comes to consider the magnetic properties of transition-metal compounds. However, we return now to our main enquiry concerning the effects of crystal fields upon these wavefunctions.

## 3.7  Crystal-Field Splitting of Free-Ion *D* Terms

Term wavefunctions describe the behaviour of several electrons in a free ion coupled together by the electrostatic Coulomb interactions. The angular parts of term wavefunctions are determined by the theory of angular momentum as are the angular parts of one-electron wavefunctions. In particular, the angular distributions of the electron densities of many-electron wavefunctions are intimately related to those for orbitals with the same orbital angular momentum quantum number; that is,

---

**Box 3-6**

The electron distributions in term wavefunctions and orbitals may be the same or complementary, as shown below

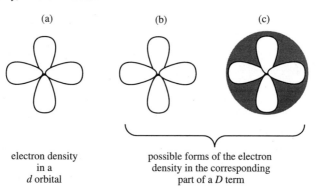

(a)          (b)          (c)

electron density
in a
*d* orbital

possible forms of the electron
density in the corresponding
part of a *D* term

The identity of (b) with (a) is obvious. In (c), a section of the density is shown to take the form of a spherical density from which a density of the form (b) has been subtracted. Alternatively, (c) may be viewed as a distribution of *positive* charge in the form of (b). Whether components of a *D* term take the form (b) or (c) depends upon the number of electrons described by the many-electron wavefunction.

---

when $L = l$. So the shapes of electron densities for the five members of a $D$ term are closely related to the shapes of the densities in the five $d$ orbitals.

In an octahedral crystal field, for example, these electron densities acquire different energies in exactly the same way as do those of the $d$-orbital densities. We find, therefore, that a free-ion $D$ term splits into $T_{2g}$ and $E_g$ terms in an octahedral environment. The symbols $T_{2g}$ and $E_g$ have the same meanings as $t_{2g}$ and $e_g$, discussed in Section 3.2, except that we use upper-case letters to indicate that, like their parent free-ion $D$ term, they are generally many-electron wavefunctions. Of course we must remember that a term is properly described by both orbital- and spin-quantum numbers. So we more properly conclude that a free-ion term $^{2S+1}D$ splits into $^{2S+1}E_g + {}^{2S+1}T_{2g}$ in octahedral symmetry. Notice that the crystal-field splitting has no effect upon the spin-degeneracy. This is because the crystal field is defined completely by its ordinary $(x, y, z)$ spatial functionality: *the crystal field has no spin properties.*

Consider, for a change, the ground term of the $d^4$ configuration. A quick way of determining the free-ion ground term (and *only* the ground term) is as follows. Hund's rules require that the ground term be of maximum spin multiplicity. We draw a set of five boxes to represent the five $m_l$ values for $d$ orbitals placing the four electrons in parallel

$$m_l = \quad 2 \quad\quad 1 \quad\quad 0 \quad\quad -1 \quad\quad -2$$

| ↑ | ↑ | ↑ | ↑ | |
|---|---|---|---|---|

spin formation. Pauli's exclusion principle then requires that these electrons occupy different $d$ orbitals (different $m_l$ values). The total $z$ component of orbital angular momentum $M_L$ is just the sum of the individual $m_l$ values, $2 + 1 + 0 - 1 = 2$. All other possible arrangements of four parallel electrons within the set of five boxes yield $M_L$ values in this way which range from 2 to $-2$. Hence $L$ for the set is 2, *i.e.* $D$. The total spin quantum number $M_S$ is $\sum (m_s)_i = 4 \times 1/2 = 2$ and hence $S = 2$. The ground term is therefore $^5D$. In an octahedral field the $^5D$ term splits to give $^5T_{2g} + {}^5E_g$ octahedral-field terms. (They are still called 'terms' because they are all many-electron wavefunctions).

Finally, consider also the case of the $d^1$ configuration. The ground (and in this case, only) term, worked out as above, is $^2D$. The reader may object that for $d^1$, we should not use upper-case labels because we are dealing with a single electron rather than with a many-electron wavefunction. But we can do so, because the word 'many' subsumes the particular case of 'one'. To say that the $^2D$ term of $d^1$ splits into $^2T_{2g} + {}^2E_g$ in an octahedral crystal field is merely to put our knowledge of the $d \rightarrow t_{2g} + e_g$ splitting onto a uniform basis so that we can compare all $d^n$ configurations. There are other advantages too, as we shall see in Section 3.10.

## 3.8 Crystal-Field Splitting of Free-Ion *F* Terms

The ground term of the $d^2$ configuration is $^3F$. That of $d^8$ is also $^3F$. Those of $d^3$ and $d^7$ are $^4F$. We shall discuss these patterns in Section 3.10. For the moment, we only note the common occurrence of $F$ terms and ask how they split in an octahedral crystal field. As for the case of the $D$ term above, which splits like the $d$ orbitals because the angular parts of their electron distributions are related, an $F$ term splits up like a set of $f$ orbital electron densities. A set of real $f$ orbitals is shown in Fig. 3-13. Note how they comprise three subsets. One set of three orbitals has major lobes directed along the cartesian $x$ or $y$ or $z$ axes. Another set comprises three orbitals, each formed by a pair of 'clover-leaf' shapes, concentrated about two of the three cartesian planes. The third set comprises just one member, with lobes directed equally to all eight corners of an inscribing cube. In the free ion, of course, all seven $f$ orbitals are degenerate. In an octahedral crystal field, however, the

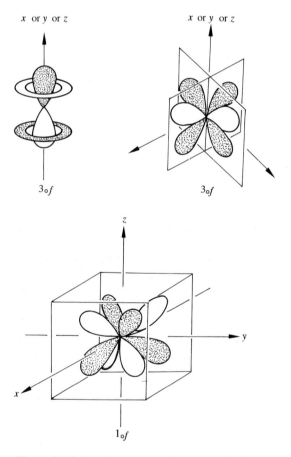

**Figure 3-13.** Angular forms of the seven $f$ orbitals.

electron densities associated with these orbitals are differentially repelled. Those for the first set are repelled most because the orbital lobes are pointing directly at the various ligands. Those of the second set are equivalent amongst themselves and are repelled less since they point between the ligands. The unique $f$-orbital density is the least repelled because each lobe is even further from the ligand sites. All this is summarized in Fig. 3-14, showing the splitting of $f$ orbitals in an octahedral field as Fig. 3-4 showed the equivalent splitting for $d$ orbitals. The orbital subsets are labelled $t_{1u}$, $t_{2u}$, $a_{2u}$ respectively for the three sets described above. All $f$ orbitals are odd (*ungerade*) under inversion through the centre of symmetry possessed by the octahedral field and so are labelled with the $u$ subscript here.

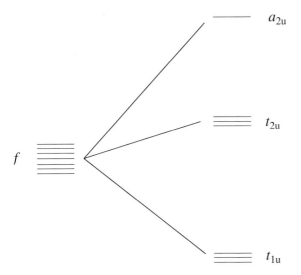

**Figure 3-14.** Splitting of $f$ orbitals in octahedral symmetry.

$F$ terms split analogously but, because we are here considering $F$ terms arising from $d^n$ configurations (we could, but don't, consider $F$ terms arising from $f^n$ configurations, by the way), the many-electron wavefunctions are built from products of $d$ orbitals of $g$ symmetry. Hence, the octahedral-field terms arising are necessarily of $g$ symmetry and so we get the result $F \rightarrow T_{1g} + T_{2g} + A_{2g}$. Another consequence of the difference between $F$ terms built from $d$-orbital products versus $f$ orbitals is that the sign or sense of the splitting depends upon the number of $d$ electrons and may or may not be the same as that shown for the $f$ orbitals in Fig. 3-14. There are various ways of determining whether the $^3F$ term from $d^2$ splits with a $^3T_{1g}$ term lowest in energy, or with the $^3A_{2g}$ term lowest. Although we have not prepared the groundwork in the present text to describe the more direct of these routes, we *are* able to decide the issue, however, by reference to the strong-field scheme. Thus, in the case of $d^2$, for example, we know from the discussion in Section 3.5 that the lowest energy arrangement in an octahedral field is orbitally three-fold degenerate. This establishes the term splitting for the $d^2$ case as that shown in Fig. 3-15. Note, once again, that both free-ion and octahedral-field terms are all spin triplets.

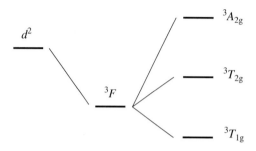

**Figure 3-15.** Splitting of the $^3F$ term arising from the $d^2$ configuration in octahedral symmetry.

## 3.9 Free-Ion S and P Terms in Crystal Fields

An $S$ term, like an $s$ orbital, is non-degenerate. Therefore, while the effect of a crystal field (of *any* symmetry) will be to shift its energy, there can be no question of its splitting. The ground term for the $d^5$ configuration is $^6S$. In an octahedral crystal field, this is relabelled $^6A_{1g}$; in tetrahedral symmetry, lacking a centre of inversion, it is labelled $^6A_1$.

The three $p$ orbitals are directed along the three cartesian axes and so, in an octahedral crystal field, suffer equal repulsion from point charges sited on those axes. The energies of the three $p$ orbitals, therefore, remain degenerate. Similarly, a free-ion $P$ term remains unsplit in octahedral or tetrahedral crystal fields and is labelled $T_{1g}$ or $T_1$ respectively.

The only spin-triplets arising from the $d^2$ configuration are $^3F$ (ground) and $^3P$. The effects of an octahedral or tetrahedral field upon these two terms are summarized in Fig. 3-16.

We note that three spin-allowed electronic transitions should be observed in the '$d-d$' spectrum in each case. We have, thus, arrived at the same point established in Section 3.5. This time, however, we have used the so-called 'weak-field' approach. Recall that the adjectives 'strong-field' and 'weak-field' refer to the magnitude of the crystal-field effect compared with the interelectron repulsion energies represented by the Coulomb term in the crystal-field Hamiltonian,

$$H_{CF} = \sum_i^n H_{H-like}(i) + \sum_{i<j}^n \frac{e^2}{r_{ij}} + V_{CF} \tag{3.13}$$

where $V_{CF}$ is the so-called crystal-field potential. Its a question of the order in which we consider these two effects. In the weak-field scheme, we begin with the free-ion terms left after the Coulomb interactions and then consider a weak crystal-field potential. In the strong-field approach, we begin with the strong-field configuration,

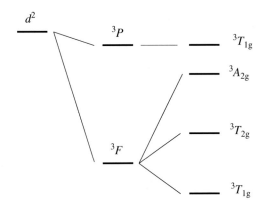

Figure 3-16. Octahedral field spin-triplet terms arising for $d^2$.

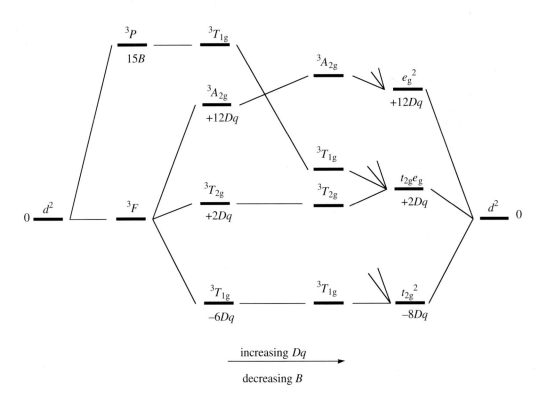

**Figure 3-17.** A partial correlation diagram for $d^2$ showing weak- (left) and strong-field (right) limits for spin-triplet terms.

recognizing the crystal-field splitting of the $d$ orbitals, and then take account of the interaction repulsion energies. Of course, the true state of affairs in any real metal complex lies somewhere inbetween the weak-field and strong-field limits. The continuum between these limits is shown semi-quantitatively by the (partial) *correlation diagram* in Fig. 3-17. On the left are shown the effects of the progressive introduction of the components of the crystal-field Hamiltonian, $H_{CF}$ in the order given in Eq. (3.13) – the weak-field approach. On the right, these contributions are taken in the order $H_{H\text{-like}}$, $V_{CF}$ and Coulomb. The term energies inbetween the weak- and strong-field limits vary across the abscissa with respect to the relative magnitudes of the Coulomb and crystal-field contributions. Unjoined lines in the diagram refer to spin-singlet terms which are otherwise omitted for simplicity. The diagram indicates that, for $d^2$ in octahedral symmetry, the ground $^3T_{1g}$ term from $^3F$ correlates with the strong-field configuration $t_{2g}^2$; that the $^3T_{2g}$ correlates with $t_{2g}^1 e_g^1$; $^3A_{2g}$ with $e_g^2$; and that the $^3T_{1g}$ term arising from the $^3P$ free-ion term – labelled $^3T_{1g}(P)$ as opposed to $^3T_{1g}(F)$ – correlates with $t_{2g}^1 e_g^1$.

We have noted how the relative term energies vary across the diagram. In particular, while the energy separations between the three strong-field configurations on the right are $\Delta_{oct}$ or $10Dq$, the energy separations between the $^3T_{1g}$, $^3T_{2g}$ and $^3A_{2g}$ terms arising from $^3F$ on the extreme left (weak-field *limit*) are $8Dq$ and $10Dq$, as shown. So while the $^3T_{2g}$ and $^3A_{2g}$ terms stay $10Dq$ apart right across the diagram, the energy separation, $^3T_{1g}(F) \leftrightarrow ^3T_{2g}$, varies from $8Dq$ to $10Dq$ as the crystal field is increased (note that this change is over and above that due to the variation in the value of $Dq$ itself). This effect is often represented by the diagram in Fig. 3-18. On the left side of the diagram, the energy separation between the $^3P$ and $^3F$ terms of the free ion is denoted as $15B$, where $B$ is a parametric measure of the magnitude of the Coulomb interaction – in effect, $B$ is for interelectron repulsion what $Dq$ is for the crystal field. It is not appropriate here for us to enquire further into the choice of this, seemingly odd, symbolism.[*] Moving one step to the right in Fig. 3-18, we see the crystal-field term energies in the weak-field limit. Note that the unsplit $^3P$ term does not shift either. This is because $V_{CF}$ is defined with respect to a barycentre rule so that, as elsewhere in both Fig. 3-17 and 3-18, the crystal-field effects ignore the overall shifts of Fig. 3-3 and refer just to *splitting* energies. Then, as we move to the right of Fig. 3-18, the energies of *both* $^3T_{1g}$ terms shift by some small amount $xDq$. This is because these wavefunctions possess the same symmetry and can mix together (in the same way that two $s$ type molecular orbitals in a diatomic molecule can mix together). As they mix, their energies change such that the energy of the higher one [$^3T_{1g}(P)$] increases while that of the lower one [$^3T_{1g}(F)$] decreases by the same amount. From Fig. 3-17, we know that the $^3T_{1g}(F)$ takes an energy $8Dq$ less than $^3T_{2g}$ in the strong-field *limit*. Hence, $x$ in Fig. 3-17 lies between 0 and 2. Again, the actual magnitude of $x$ in any real system is a measure of the relative magnitudes of the crystal-field and interelectron repulsion effects. The

---

[*] Except to note that the occurrence of the coefficients 15 and 10 in $15B$ and $10Dq$ obviate the need for fractions here or elsewhere in crystal-field theory: thus, they are there for reasons of convenience and definition only.

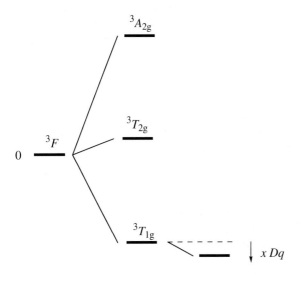

$^3P$ — 15B

$^3T_{1g}$

$\uparrow x\,Dq$

$^3A_{2g}$

$^3T_{2g}$

$^3F$ — 0

$^3T_{1g}$

$\downarrow x\,Dq$

**Figure 3-18.** Interaction between $^3T_{1g}(F)$ and $^3T_{1g}(P)$ terms.

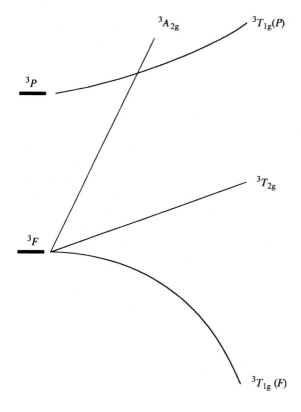

$^3A_{2g}$

$^3T_{1g}(P)$

$^3P$

$^3T_{2g}$

$^3F$

$^3T_{1g}(F)$

**Figure 3-19.** Schematic representation of the energy variations of the spin-triplets of $d^2$.

changes across Fig. 3-18 are not discontinuous, of course, but smooth and may be redrawn as in Fig. 3-19. The energies of the $^3A_{2g}$ and $^3T_{2g}$ terms vary linearly with $Dq$ with constant *relative* slopes of $10Dq$. Those for the two $^3T_{1g}$ terms curve away from one another (as if they repel each other). We shall refer to this diagram again in Section 3.11.

# 3.10 Splitting Patterns for $d^n$ Ground Terms

A complete description of the effects of a crystal field upon a $d^n$ ion would include similar analysis of the behaviours of all terms arising for that $d^n$ configuration. Box 3-7 summarizes the case for $d^2$, and in Box 3-8, we illustrate a method of using Fig. 3-19 to determine $Dq$ and $B$ values from real spectra.

Much useful understanding of the processes of crystal-field theory, however, can be had from a study of just the free-ion ground terms. Application of the simple process in Section 3.7 identifies the ground terms for $d^n$ free ions as $^2D$, $^3F$, $^4F$, $^5D$, $^6S$, $^5D$, $^4F$, $^3F$, $^2D$ for $n = 1$ to 9 respectively. There are some interesting patterns in this series. Firstly, note that the ground terms for $d^{10-n}$ configurations are the same as for $d^n$. This is because a $d^{10-n}$ configuration of electrons behaves in many ways like a $d^n$ configuration of holes. There is no question of any inversion here, however, because two holes repel each other just as do two electrons. This symmetry between $d^{10-n}$ and $d^n$ configurations extends to all terms. It does not apply to the absolute energies of any of the free-ion terms, however, for the mutual shielding for $10-n$ electrons is not the same as for $n$ electrons. Secondly, if spin is ignored, we also observe a symmetry at the 1/4 and 3/4 periods[*]. Thus, the orbital (spatial) parts of the ground terms for $d^{5+n}$ are the same as for $d^n$, as also are those for $d^{10-n}$ like $d^{5-n}$. The reason for this can be seen in two ways. With the restriction to the ground term (*i.e.* of maximum spin multiplicity), $d^{5-n}$ involves $n$ holes in the half-shell of five $d$ orbitals. The number and manner of the arrangement of, say, three electrons amongst the five $d$ orbitals is the same as the number and manner of two holes: similarly for, say, $d^7$ and $d^8$ configurations. The other way to look at it, with the same restriction of maximum spin multiplicity, is as follows: The configuration $d^{5+n}$ (maximum spin) is like that for $d^n$ plus a filled half-shell. Thus, for $n = 1$:

Ignoring spin, the ground term for $d^{5+n}$ is the same as for $d^n$. But, from the first symmetry, as described above, that $d^{10-n}$ is like $d^n$, this implies that $d^{5-n}$ is like $d^n$ for the ground term. By way of emphasis: the symmetry at the half-period – $d^{10-n}$

---

[*] The 1/4, 1/2, and 3/4 periods occur between $d^3$ and $d^4$; at $d^5$; and between $d^7$ and $d^8$, respectively.

**Box 3-7**

Supplementing the knowledge we have so far about the splitting of $S$, $P$, $D$ and $F$ terms in octahedral fields is the fact that a $G$ term gives rise to $A_{1g} + E_g + T_{2g} + T_{1g}$ crystal-field components. The diagram in the figure below – a so-called 'Tanabe-Sugano' diagram - shows the energies of all octahedral-field terms arising from the $d^2$ configuration as functions of $Dq$ and $B$. By convention, energy levels are plotted with respect to the ground term. Curvatures in these variations result in part from mixing between terms of the same kind (same spin and spatial symmetry) and in part because this method of presentation has the ground term along the abscissa.

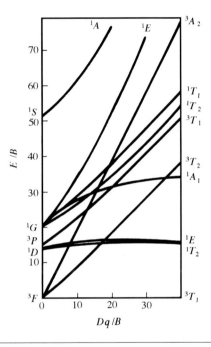

like $d^n$ – applies to *all* terms of these configurations while the symmetries at the 1/4 and 3/4 periods apply *only* to the ground terms.

The effects of an octahedral crystal field upon each of the ground terms is shown in Fig. 3-20. This diagram was constructed as follows. From our previous discussions, $D \rightarrow T_{2g} + E_g$, $F \rightarrow T_{1g} + T_{2g} + A_{2g}$ (with $T_{2g}$ always in the middle), and $S \rightarrow A_{1g}$. For the $d^1$ configuration, $^2T_{2g}$ lies lower than $^2E_g$. The symmetries we observed for the ground terms of the free-ion terms at the 1/4, 1/2 and 3/4 periods also apply for the crystal-field splittings except that, while two electrons (or holes) always repel one another for the Coulomb contribution, electrons are repelled by the crystal field but holes are attracted. So these symmetries give the *same* results for the *two-electron* Coulomb operator but *inverse* results for the *one-electron* crystal-field operator. Applying these rules, the $^2D$ of $d^9$ yields $^2T_{2g}$ and $^2E_g$ terms in the opposite energy order to those of $d^1$, *i.e.* $^2E_g$ lower than $^2T_{2g}$. The $^5D$ term of $d^4$

---

**Box 3-8**

The figure below abstracts just the spin-triplet part of the Tanabe-Sugano diagram in the previous box. Suppose we have recorded the electronic '$d-d$' spectrum of $[V(H_2O)_6]^{2+}$ and identified two out of the three possible 'spin-allowed' (triplet–triplet) bands at energies 17,200 cm$^{-1}$ and 25,600 cm$^{-1}$

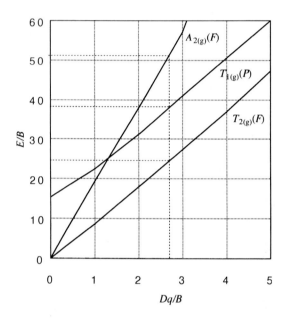

The question arises 'to which of the transitions $^3T_{1g} \rightarrow {}^3T_{2g}, \rightarrow {}^3A_{2g}, \rightarrow {}^3T_{1g}(P)$ do these bands correspond?'.

We proceed by determining at what value of $Dq/B$ on the abscissa is the ratio $256/172 = 1.49$ reproduced by the plot: this is shown by the vertical broken line in the figure . Then we construct the horizontal broken lines to meet, as shown. For $v_a$ we find $E/B$ on the ordinate as 25.9, and for $v_b$ we get $E/B = 38.7$. From either we find, therefore, that $B = 665$ cm$^{-1}$. On referring back to the vertical line we thus find $10Dq = 18,600$ cm$^{-1}$. At the same time we have established that $v_a$ corresponds to the transition $\rightarrow {}^3T_{(2g)}(F)$ and $v_b$ to $\rightarrow {}^3T_{1g}(P)$. The transition $\rightarrow {}^3A_{2g}(F)$ is predicted to lie at ca. 36,000 cm$^{-1}$.

---

also splits inversely with respect to the $^2_2D$ term of $d^1$; the $^5D$ term of $d^6$ splits inversely with respect to $d^9$ and hence the same way round as for the $^2D$ of $d^1$.

Turning now to the $F$ terms. We have previously established that the ground term of $d^2$ in octahedral symmetry is $^3T_{1g}$. Therefore, by the symmetry in the 1/2 period, that for $d^8$ is $^3A_{2g}$ and by the symmetries in the 1/4 and 3/4 periods, the ground terms for $d^3$ and $d^7$ are $^4A_{2g}$ and $^4T_{1g}$ respectively.

Finally, the $^6S$ term of $d^5$ does not split and is labelled $^6A_{1g}$: $d^5$ is its own hole-equivalent. While there seems to be a lot to remember in Fig. 3-20, all one needs to commit to memory is the ordering for, say, $d^1$ and $d^2$. The hole formalisms –

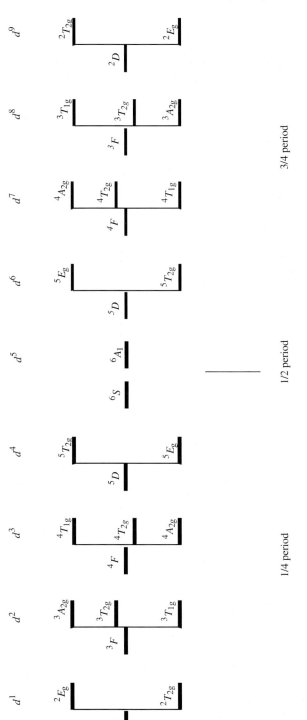

**Figure 3-20.** The symmetrical pattern of ground term splittings in octahedral symmetry.

inversions at 1/4, 1/2 and 3/4 periods – do the rest. Lastly, consider tetrahedral symmetry. In Section 3.3, we saw how the $d$ orbitals split with ($d_{xz}$, $d_{yz}$, $d_{xy}$) higher in energy than ($d_{z^2}$, $d_{x^2-y^2}$); in other words, inverted with respected to the octahedral- field splitting. One reason why some prefer to write (3.2) as $\Delta_{\text{tet}} = -{}^4/_9\Delta_{\text{oct}}$ is to indicate that 'the tetrahedral crystal field is (4/9 times) the negative of the octahedral field'. This remark is not intended to confuse but merely to provide background to our assertion now that *all* splitting diagrams in Fig. 3-20 are inverted for tetrahedral crystal fields. Of course, the $g$ subscripts are omitted in the tetrahedral case.

## 3.11 Orgel Diagrams

The information in Section 3.9 and Section 3.10, referring to the crystal-field splittings of ground terms and all terms of the same spin multiplicity, can be very neatly encapsulated within two famous diagrams due to Orgel. Somewhat analogous,

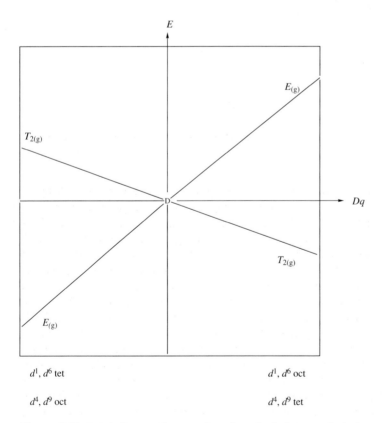

$d^1$, $d^6$ tet          $d^1$, $d^6$ oct

$d^4$, $d^9$ oct          $d^4$, $d^9$ tet

**Figure 3-21.** Orgel diagram for energies of octahedral or tetrahedral terms arising from free-ion $D$ terms.

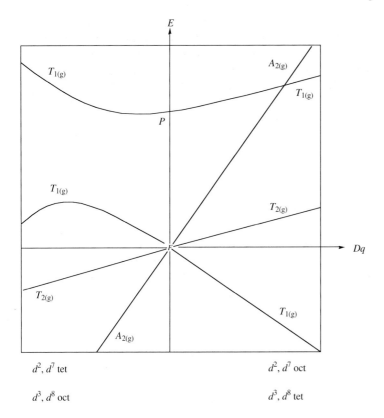

**Figure 3-22.** Orgel diagram for energies of octahedral or tetrahedral terms arising from free-ion $F$ or $P$ terms.

and more comprehensive, diagrams for the effects of crystal fields on *all* terms arising from a given $d^n$ configuration are called Tanabe-Sugano diagrams: an example was given in a Box 3-7 above. The Orgel diagrams are shown in Fig. 3-21 and 3.22. They refer to the octahedral- *or* tetrahedral-field splitting of ions with $D$ or $F$ ground terms respectively. Ions with $d^1$, $d^4$, $d^6$ or $d^9$ configurations possess only one term of maximum spin multiplicity – $^2D$ or $^5D$, as appropriate. The crystal-field $T_{2(g)}$ and $E_{(g)}$ terms that arise are unique in each case and there are no interactions between terms of the same symmetry as in Section 3.9. Thus, the energies of these $T_{2(g)}$ and $E_{(g)}$ terms vary linearly with $10Dq$ as shown in Fig. 3-21. The two sides of the figure depict the situations with either a $T_{2(g)}$ term or an $E_{(g)}$ term being lowest in energy. The slopes of the lines arise from the barycentre rules so that the energy of the more degenerate $T_{2(g)}$ term slopes more gently than that of the $E_{(g)}$ term. This corresponds to the energies $+6Dq$ and $-4Dq$ in Fig. 3-4, for example. The inversions in the 1/4, 1/2 and 3/4 periods as well as that on changing octahedral for tetrahedral symmetry are all accounted for in the abscissa labels. Recall our remark in Section 3.3 that we sometimes refer to the orbital triplet-singlet splitting in *both* octahedral and tetrahedral symmetry as $10Dq$, rather than

$\Delta_{oct}$ or $\Delta_{tet}$. The context should remove ambiguity. On the other hand, that very ambiguity means that Orgel's diagrams are always drawn with $Dq$ varying along the abscissae, regardless of octahedral or tetrahedral symmetry. Equivalently, the $g$ subscripts are enclosed in parentheses here so that these diagrams serve both geometries.

Similar rules of construction have been followed in Fig. 3-22. Here, the higher lying $^3P$ or $^4P$ terms (as appropriate) are included. Curvatures due to the repeated (mixing) $T_{1(g)}$ terms are shown, corresponding to the representation we have discussed in Fig. 3-19. Note, however, that the interaction between terms of the same symmetry is essentially inversely proportional to their energy separation: the closer they are in energy the more they mix and 'repel' one another. Because of the various inversions we have described, the highest energy term arising from an $F$ term can be the $T_{1(g)}$. In this case, the $T_{1g}(F)$ and $T_{1g}(P)$ would be brought into close proximity and might be expected to cross with increasing crystal-field strength. Their interaction prevents their crossing (see also the 'non-crossing rule') and the curvatures displayed in the Orgel diagrams in these circumstances are very great.

## 3.12 Concluding Remarks

In the next chapter we look at the intensities of '$d-d$' electronic transitions. We shall see that transitions between terms of the same spin-multiplicity are much more intense than those involving a change of spin. It is for this reason that our focus in the present chapter has been on the former. We have seen that for $d^1$, $d^4$, $d^6$ and $d^9$ configurations in octahedral or tetrahedral environments, there is only one so-called 'spin-allowed' transition. For $d^2$, $d^3$, $d^7$ and $d^8$ configurations there are three. There are none at all for ions with the $d^5$ configuration. The energies of the transitions in the former group depend only upon the strength of the crystal field that is, upon the extent to which the metal electrons seek to avoid the ligands. On the other hand, the transition energies for ions in the second group are functions of the extents to which the $d$ electrons avoid each other *and* the external environment. Quantitative considerations of both contributions yield much chemical information, as we shall see in Chapter 6.

Our discussions throughout have been based on the crystal-field model. That is to say, the physical origin of the splittings has been ascribed to the repulsion of the $d$ electrons by negatively charged ligands. Even at this early stage in our exposition, it is well to realize that the same qualitative picture would emerge whatever the detailed physical origin of these splittings is, provided it is such that the $d$ electrons closest to ligand electron density are most raised in energy. In one sense, of course, that will suffice as a definition of repulsion, but there is much more to be said yet about the origin of these effects.

# Suggestions for further reading

1. B.N. Figgis, *Introduction to Ligand Fields*, Wiley, New York, **1966.**
2. F.A. Cotton, *Chemical Applications of Group Theory*, 3rd ed., Wiley, New York, **1990.**
   – These references describe the material of chapter more fully and at a somewhat more technical level; they provide good insight.
3. C.J. Ballhausen, *Introduction to Ligand-Field Theory*, McGraw-Hill, New York, **1962.**
4. J.S. Griffith, *Theory of Transition Metal Ions*, Cambridge University Press, Cambridge, **1961.**
   – These two books delve much more deeply, and at a much more mathematical level, into this subject.
5. M. Gerloch, *Orbitals, Terms and States*, Wiley, New York, **1986.**
   – This book describes these quantities for atoms and linear molecules thoroughly but at a not-too-difficult mathematical level.

# 4 The Intensities of '$d-d$' Spectra

## 4.1 Transition moments

We shouldn't think of the absorption of light by a molecule in an anthropomorphic way as if the molecule takes in the light and looks for something to do with it! The various electronic states in a molecule correspond to discrete arrangements of the electrons in the molecule. When we refer to a molecule as 'being in its ground state or one of its excited states' we mean that it has such-and-such an electronic arrangement with such-and-such an energy. The absorption of light by a molecule is an interaction of the light with the molecule. The effect upon the molecule is that its electrons rearrange from one state to another. We talk of a spectroscopic *transition* from one electronic state to another. As mentioned in Section 2.1, the lifetime of excited electronic states is normally very short – say $10^{-18}$ sec – and the electrons spontaneously rearrange to the ground state with re-emission of the light. These transitions only take place when the frequency (or energy, *via* the relationship $E = h\nu$) of the light matches the energy difference between the two states involved in the transition. The probability of a given transition taking place depends upon the initial and final states and upon the transition-inducing properties of the light; where we start, where we finish, and what causes it. We define a *transition moment*, $Q$, as the integral,

$$Q = \int \psi^*_{gd} \cdot (\text{light operator}) \cdot \psi_{ex} \, d\tau$$
$$= < \psi_{gd} \mid \text{light operator} \mid \psi_{ex}> \tag{4.1}$$

where $\psi_{gd}$ and $\psi_{ex}$ represent the ground and excited state wavefunctions. The intensity, $I$, of this transition – that is, its probability – is given by the square of the transition moment,

$$I \propto Q^2 \tag{4.2}$$

or by $I \propto Q^*Q$ when $Q$ is complex.

Now we need to know a little about the nature of the transition-inducing light. Light is a longitudinally propagating transverse oscillating electromagnetic field. As shown in Fig. 4-1, the electric and magnetic oscillations are at right angles to one another and to the direction of propagation. The arrows in the diagram emphasise that at any point along the propagating beam, there are electric and

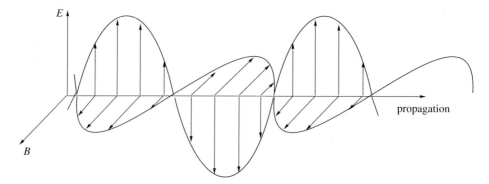

**Figure 4-1.** Light as transverse electric ($E$) and magnetic ($B$) oscillating fields normal to the propagation direction.

magnetic fields of definite sign or parity (except, trivially, where they are zero). In short, these electric and magnetic fields are *dipolar*. We write our operator for the light in Eq. (4.1), therefore, as a combination of electric and magnetic dipole operators. However, essentially because of the small size of molecules (or of the chromophoric part of the molecule – meaning that part undergoing electronic transitions) compared with the wavelength of the light typically used, the ability of the *magnetic* dipole to cause transitions is very slight – typically $10^{-4}$ times that of the electric dipole. Accordingly, we need not consider so-called magnetic-dipole transitions any further and write transition moments as

$$Q = < \psi_{gd}|er|\psi_{ex} > \qquad (4.3)$$

where the electric-dipole operator $er$ comprises the electronic charge, $e$, and the radius vector, $r$, describing the orientation of the unit light electric dipole.

## 4.2 Selection Rules

Suppose each state wavefunction in Eq. (4.3) can be written as a simple product of space-only and spin-only parts: then $Q$ is given by

$$Q = < \psi_{space} \, \psi_{spin} \mid er \mid \psi'_{space} \, \psi'_{spin} > \qquad (4.4)$$

where we have changed notation by using primes for the excited state for typographical reasons. The space parts of the wavefunctions depend only upon ordinary $(x, y, z)$ space while the spin parts are functions only of spin space. The operator $er$ is a function only of ordinary space and therefore does nothing to any spin function. Regrouping variables within the integral, $Q$, we find Eq. (4.5).

$$Q = < \psi_{\text{space}} \mid e\mathbf{r} \mid \psi'_{\text{space}} > < \psi_{\text{spin}} \mid \psi'_{\text{spin}} > \qquad (4.5)$$

Regardless of the nature of the space parts, $Q$ vanishes if $\psi_{\text{spin}} \neq \psi'_{\text{spin}}$. If $Q$ vanishes, so does $I$. Thus we have the so-called *spin-selection rule* which denies the possibility of an electronic transition between states of different spin-multiplicity and we write $\Delta S = 0$ for spin-allowed transitions. Expressed in different words, transitions between states of different spin are not allowed because light has no spin properties and cannot, therefore, change the spin.

Now consider a transition between states of the same spin. The 'spin overlap integral', $< \psi_{\text{spin}} \mid \psi'_{\text{spin}}>$, in Eq. (4.5) is non-zero: if all relevant functions are normalized, it is unity. So we turn our attention to the space part of $Q$ (Eq. 4.6).

$$Q = < \psi_{\text{space}} \mid e\mathbf{r} \mid \psi'_{\text{space}} > ; \text{space part} \qquad (4.6)$$

Generally, such integrals are calculable only with great difficulty because we rarely know the exact forms of the wavefunctions $\psi$. However, a great deal can be established simply by considerations of symmetry. We rely on a general mathematical concept that integrals of odd functions vanish. An odd function is one with two parts (or, in general, two sets of parts) which are identical in shape but opposite in sign. An even function is one with parts which are identical in shape and sign. Figure 4-2 shows some examples. A particular, though not unique, subset of such functions are those which are odd or even with respect to inversion through a centre of symmetry. Examples we are familiar with are the $d$ and $p$ orbitals, being even and odd respectively.

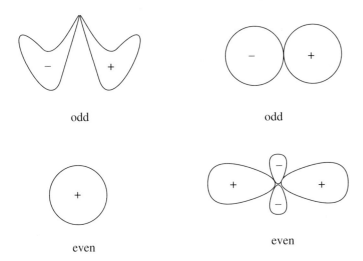

**Figure 4-2.** Examples of odd and even functions.

Let us enquire about the electric dipole transition moment between two $d$ orbitals as expressed in Eq. (4.7).

$$Q_{space} = <d \mid er \mid d'>$$ (4.7)

The $d$ orbitals are centrosymmetric and are of $g$ symmetry. The light operator, being dipolar, is of $u$ symmetry. The symmetry of the whole function under the integral sign in (4.7) – that is, for the product $d \cdot r \cdot d$ – is $g \times u \times g$, namely $u$. The integral over all volumes of a $u$ function vanishes identically. Since $Q$ in (4.7) then vanishes, so does the intensity $I$. In short, '$d–d$' transitions are disallowed.

This is an example of the *parity rule*, also known as *Laporte's rule*: transitions between orbitals or states of different parity are forbidden. It means that all $d–d$, $p–p$, $s–s$ and $f–f$ orbital transitions, for example, are forbidden. Actually, the disallowedness of these transitions is underscored by another selection rule. The photons of light possess one unit of angular momentum (a spatial property). The absorption of a photon during the course of a transition vectorially adds exactly one unit of angular momentum to the molecule. This means that the two orbitals in a transition must have orbital angular momenta that differ by exactly one unit: $\Delta l = \pm 1$. Note also that the orbital selection rule is $\Delta l = \pm 1$ and NOT $\Delta L = \pm 1$. The absorption of one photon of light leads to the rearrangement of one electron, that is, we are dealing with a one-electron property here – hence the lower-case $l$. So transitions $s \rightarrow p$, $p \rightarrow d$, $d \rightarrow f$ are allowed but not $d \rightarrow d$ etc.: the rule also forbids transitions like $s \rightarrow d$ or $p \rightarrow f$ which would be allowed by the parity rule.

## 4.3 'Violation' of the Selection Rules

These selection rules appear to predict that transition-metal complex spectra will have no '$d–d$' bands. But, of course, they do! The rules are strictly absolute: they may not be violated at all. Their apparent violation derives from the nature of the electronic states involved in any transition. At the beginning of Section 4.2, we *supposed* that the relevant wavefunctions could be factorized into space-only and spin-only parts. While that is nearly true, it isn't exactly so. The magnetic interaction which couples the orbital- and spin-angular momenta of electrons, that we call 'spin-orbit coupling', means that neither spin nor orbital properties are constant with time. In the same way that the electrostatic coupling between electrons causes the angular momenta of each electron to vary with time in favour of constant spin and orbital momenta for the electrons *as a group*, the magnetic interaction yields only a *total* angular momentum that is constant. For a single electron, we write the total angular momentum $j$ as the vector sum in (4.8)

$$j = l + s$$ (4.8)

and for a group of electrons, we write it as in (4.9).

$$J = L + S$$ (4.9)

A term label like $^3F$, for example, is thus no longer *strictly* meaningful for it implies constant spin- and orbital angular momentum properties ($S = 1$, $L = 3$). One consequence of spin-orbit coupling is a 'scrambling' of the two kinds of angular momentum. So a nominal $^3X$ term may really more properly be described as a mixture of terms of different spin-multiplicity as, for example, in Eq. (4.10).

$$`^3X` = {}^3X + a^1Y$$
$$\text{or} \qquad `^4Q` = {}^4Q + b^2R \text{ , etc} \qquad\qquad (4.10)$$

The mixing coefficients $a$ and $b$ in (4.10) depend upon the efficiency of the spin-orbit coupling process, parameterized by the so-called spin-orbit coupling coefficient $\lambda$ (or $\zeta$ for a single electron). As $\lambda \to 0$, so also do $a$ or $b$. Spin-orbit coupling effects, especially for the first period transition elements, are rather small compared with either Coulomb or crystal-field effects, so the mixing coefficients $a$ or $b$ are small. However, insofar that they are non-zero, we might write a transition moment as in Eq. (4.11).

$$Q = < `^3X` \mid er \mid `^1W` >$$
$$= < {}^3X + a^1Y \mid er \mid {}^1W + b^3Z >$$
$$= < {}^3X \mid er \mid {}^1W > + ab < {}^1Y \mid er \mid {}^3Z > + a < {}^1Y \mid er \mid {}^1W > + b < {}^3X \mid er \mid {}^3Z >.$$
$$(4.11)$$

The first two terms in the expansion are strictly zero because of the spin selection rule, while the last two are non-zero, at least so far as the spin-selection rule is concerned. So a 'spin-forbidden' transition like this, $`^3X` \to `^1W`$, can be observed because the descriptions $^3X$ and $^1W$ are only approximate: that is why we enclose them in quotation marks. To emphasize: the spin-orbit coupling coefficients for the first row transition elements are small, the mixing coefficients $a$ and $b$ are small, and hence the intensities of these spin-forbidden transitions are very weak.

Consider now spin-allowed transitions. The parity and angular momentum selection rules forbid pure $d \leftrightarrow d$ transitions. Once again the rule is absolute. It is our description of the wavefunctions that is at fault. Suppose we enquire about a '$d$–$d$' transition in a tetrahedral complex. It might be supposed that the parity rule is inoperative here, since the tetrahedron has no centre of inversion to which the $d$ orbitals and the light operator can be symmetry classified. But, this is not at all true; for two reasons, one being empirical (which is more of an observation than a reason) and one theoretical. The empirical 'reason' is that if the parity rule were irrelevant, the intensities of '$d$–$d$' bands in tetrahedral molecules could be fully allowed and as strong as those we observe in dyes, for example. In fact, the '$d$–$d$' bands in tetrahedral species are perhaps two or three orders of magnitude weaker than many fully allowed transitions.

The theoretical reason is as follows. Although the placing of the ligands in a tetrahedral molecule does not define a centre of symmetry, the $d$ orbitals are nevertheless centrosymmetric and the light operator is still of odd parity and so $d$–$d$ transitions remain parity and orbitally ($\Delta l = \pm 1$) forbidden. It is the *nuclear* coordinates that fail to define a centre of inversion, while we are considering a transition of *electron* coordinates. However, with respect to those nuclear coordinates, no functions can be *labelled* as *g* or *u*. It is entirely possible that proper wavefunctions (solutions of the Schrödinger equation for the *molecule*) may involve mixtures of, say, *d* and *p* wavefunctions. A true wavefunction that we label '*d*' (using the quotation mark convention again) might more properly be expressed as in Eq. (4.12).

$$'d' = d + cp \tag{4.12}$$

Appropriate transition moments then take the form in Eq. (4.13).

$$
\begin{aligned}
Q &= <'d' \mid er \mid 'd'' > \\
&= < d + cp \mid er \mid d' + c'p' > \\
&= < d \mid er \mid d' > + cc' < p \mid er \mid p' > + c < p \mid er \mid d' > + c' < d \mid er \mid p' >.
\end{aligned}
\tag{4.13}
$$

The first two parts of the expression vanish exactly because of Laporte's rule, while the last two survive both parity and orbital selection rules to the extent that the mixing coefficients $c$ and $c'$ are non-zero in noncentric complexes.

Experimentally, spin-allowed '*d–d*' bands (we use the quotation marks again) are observed with intensities perhaps 100 times larger than spin-forbidden ones but still a few orders of magnitude (say, two) less intense than fully allowed transitions. This weakness of the '*d–d*' bands, alluded to in Chapter 2, is a most important pointer to the character of the *d* orbitals in transition-metal complexes. It directly implies that the admixture between *d* and *p* metal functions is small. Now a ligand function can be expressed as a sum of metal-centred orbitals also (see Box 4-1). The weakness of the '*d–d*' bands also implies that that portion of any ligand function which looks like a *p* orbital when expanded onto the metal is small also. Overall, therefore, the great extent to which '*d–d*' bands do satisfy Laporte's rule entirely supports our proposition in Chapter 2 that the *d* orbitals in Werner-type complexes are relatively well isolated (or decoupled or unmixed) from the valence shell of *s* and/or *p* functions.

Now look at octahedral complexes, or those with any other environment possessing a centre of symmetry (*e.g.* square-planar). These present a further problem. The process of 'violating' the parity rule is no longer available, for orbitals of different parity do not mix under a Hamiltonian for a centrosymmetric molecule. Here the nuclear arrangement requires the labelling of *d* functions as *g* and of *p* functions as *u*; in centrosymmetric complexes, *d* orbitals do not mix with *p* orbitals. And yet '*d–d*' transitions *are* observed in octahedral chromophores. We must turn to another mechanism. Actually this mechanism is operative for all chromophores, whether centrosymmetric or not. As we shall see, however, it is less effective than that described above and so wasn't mentioned there. For centrosymmetric systems it's the only game in town.

When discussing the origin of the large widths of '*d–d*' bands in Chapter 2, we noted that molecules are always vibrating. Some of these vibrations are such as to remove a centre of inversion. Consider just the one example in Fig. 4-3. This

**Box 4-1**

Bonding orbitals in a metal complex may be thought of as molecular orbitals built from appropriate metal and ligand functions. In the case of an M–L $\sigma$ bond orbital, $\psi_\sigma$, for example, we write

$$\psi_\sigma = c_M \, M_\sigma + c_L \, \Phi_\sigma$$

where $M_\sigma$ and $\Phi_\sigma$ are metal and ligand $\sigma$ orbitals, and $c_M$, $c_L$ are some mixing coefficients. It is usual to think of the metal orbital as centred upon (that is, expressed with respect to) the metal and of the ligand orbital as centred on the ligand. Diagramatically, this can be represented as the following.

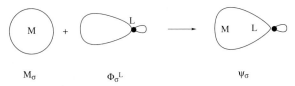

$$M_\sigma \qquad\qquad \Phi_\sigma{}^L \qquad\qquad \psi_\sigma$$

For many purposes, it is more convenient to express all functions with respect to just one origin – most usually the metal. The expansion theorem may be exploited to express any function as an (infinite) sum of convenient 'basis' functions. Here we write the function centred on the ligand as a linear combination of functions centred on the metal

$$\Phi_\sigma = a_1 s^M + a_2 p_\sigma^M + a_3 d_\sigma^M + a_4 f_\sigma^M + \ldots$$

The figure below illustrates how such a linear combination of metal orbitals (taken sequentially in the diagram for heuristic reasons) can reproduce the orbital on another centre.

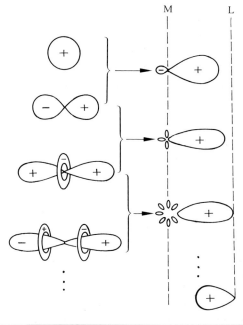

**Box 4-1** (Continued)

The point of this little diversion has been to show how a bond orbital like $\psi_\sigma$ above can, quite generally, be thought of as some linear combination of just metal-centred orbitals.

Now, ultimately, the metal $d$ orbitals become admixed with a *little* of the bond orbital $\psi$ and so within that final mixture we find both $d$ and $p$ type orbitals (others too, but we focus here just on the $d–p$ mixing). If the whole $ML_N$ complex is centrosymmetric, all such $d–p$ mixings cancel identically, but not otherwise. ***Now comes the point.*** When such $d–p$ mixing survives, was the $p$ orbital originally on the metal (*e.g.* a metal $4p$), or part of the ligand function $\Phi$? Of course, we cannot say. However, knowing that Laporte's rule is so well obeyed in practice, means that such $d–p$ mixing is small and hence, even if all the $p$ character originated on the ligand, we conclude that the metal-ligand orbital mixing is small.

vibration mode of an octahedron involves the stretching of one bond together with a simultaneous contraction of the bond on the opposite side of the metal (the other ligands make minor movements also, as shown). During the course of the vibration, therefore, the nuclear arrangement lacks any centre of inversion. At any instant during the vibration, $d–p$ mixing can occur (other than at the trivial point of zero distortion) and, in the manner of Eq. (4.13), a '$d–d$' transition can become partially allowed. The reader may object that during the other half of the vibration, the nuclear displacements will be reversed, as shown on the right in Fig. 4-3, so that the orbital mixing may change from $d_z + cp_z$ to $d_z - cp_z$. Yet this doesn't imply cancellation of contributions to the transition moment or intensity. This is because, once again, the typical period of a vibration is about $10^{-13}$ sec, compared with the approximately $10^{-18}$ sec lifetime of the electronic excitation. During the course of a

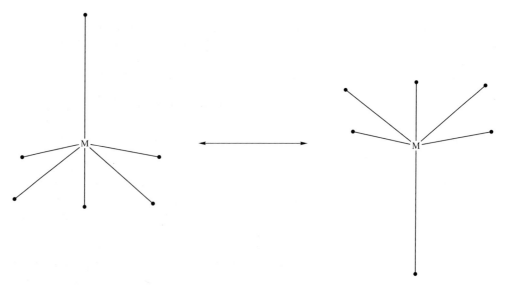

**Figure 4-3.** An *ungerade* vibration of an octahedron.

---

**Box 4-2**

Selection rules: a reminder.
A mistake often made by those new to the subject is to say that 'The Laporte rule is irrelevant for tetrahedral complexes (say) because they lack a centre of symmetry and so the concept of parity is without meaning'. This is incorrect because the light operates not upon the nuclear coordninates but upon the *electron* coordinates which, for pure $d$ or $p$ wavefunctions, have well-defined parity. The lack of a *molecular* inversion centre allows the mixing together of pure $d$ and $p$ (or $f$) orbitals: the result is the mixed parity of the orbitals and consequent non-zero transition moments. Furthermore, had the original statement been correct, we would have expected intensities of tetrahedral '$d-d$' transitions to be *fully* allowed, which they are not.

---

single vibrational cycle of the nuclei, there is time for around $10^5$ electronic excitations! Put it another way. An incident beam of light encounters an ensemble of many (perhaps $10^{20}$) molecules which appear stationary but at all possible stages of the vibrational cycle. For each encounter of light and molecule, an effectively static $d-p$ mixing is in place and a '$d-d$' absorption can occur. Overall, the 'violation' of the parity rule in vibrating octahedral chromophores is less than in static tetrahedral ones because the average degree of $d-p$ mixing is less in the dynamic environment than in the static. Typically, '$d-d$' bands for octahedral complexes are about ten to one hundred times weaker than those for tetrahedral complexes. The mechanism just described is often called 'vibronic' coupling.

## 4.4  Intensity 'Stealing'

Occasionally, some bands which might otherwise be expected to be weak are observed to be quite strong. Two examples are shown in Fig. 4-4. The first shows the electronic spectrum of a solution containing $[CoCl_4]^{2-}$ ions in nitromethane. For this $d^7$ system, we expect three spin-allowed transitions and these are observed at roughly 3500, 7000 and 14,000 cm$^{-1}$. They correspond (see Chapter 3) to the excitations $^4A_2 \rightarrow {}^4T_2$, $\rightarrow {}^4T_1(F)$ and $\rightarrow {}^4T_1(P)$ respectively. Note, however, that the band at 14,000 cm$^{-1}$ comprises several sub-maxima. In part, they are assigned to components of the $^4T_1(P)$ term that arise due to spin-orbit coupling. At least one component, however, is assigned to a spin-forbidden transition, $^4A_2 \rightarrow {}^2X(^2G)$. The details of the assignment are unimportant for us. The feature of particular interest here is that the intensity of the spin-forbidden transition is comparable with those of the spin-allowed transition. Other spin-forbidden transitions elswhere in the spectrum are very weak indeed, as generally expected. Why is this particular spin-forbidden band so strong? Well, recall the process that leads to spin-forbidden bands being seen at all. In Eq. (4.11), the allowedness of such transitions is proportional to the mixing between states of different spin angular momentum caused by spin-orbit coupling. The degree of such mixing is in turn proportional to the spin-orbit coupling coefficient which is quite small for the first row transition-metal

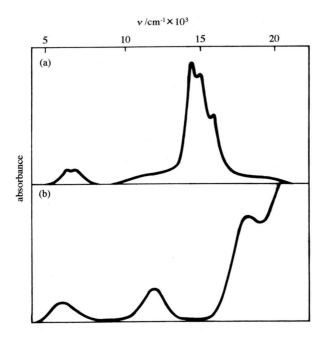

**Figure 4-4.** a) Spectrum of $[CoCl_4]^{2-}$ and b) Spectrum showing intensity stealing from a charge-transfer band at higher energy.

ions. However, it is also roughly inversely proportional to the energy separation of the states being mixed in this way. Occasionally, and the tetrachlorocobaltate(II) ion presents just such a case, spin quartet and doublet (here) terms are brought into close energetic proximity by the crystal field. The resulting scrambling of these terms by spin-orbit coupling can then be very large indeed, and it is quite possible that the true wavefunctions can approach 50:50 quartet-doublet character. Under these circumstances, the 'spin-forbidden' transition may acquire a much larger intensity than usual. Of course, the intensity gained in this way is at the expense of the intensity of the 'spin-allowed' transition since it is 'diluted', as it were, by the admixed doublet character. Between them, the $^4A_2 \rightarrow {}^4T_1(P)'$ and $^4A_2 \rightarrow {}^2X(^2G)'$ transitions possess some total intensity. As the spin-orbit induced mixing scrambles those wavefunctions, the distribution of intensity between them becomes more equal. We say that the spin-forbidden transition acquires intensity by 'stealing' from the spin-allowed band. Actually, all spin-forbidden intensities arise in this way, for the process summarized in Eq. (4.11) is really the full story, but the expression 'intensity stealing' appears to be reserved for bands with 'unusually' high intensities due to their close proximity to more intense transitions.

The second example in Fig. 4-4 shows how a (spin-allowed or spin-forbidden) band lying close to a charge transfer band may acquire unusually high intensity. We shall discuss charge-transfer bands more in Chapter 6. For the moment, we note that they involve transitions between metal $d$ orbitals and ligands, are often fully allowed and hence intense. On occasion, the symmetry of a charge transfer state may differ from that of an energetically proximate $d$ state in a way that can be matched by a molecular vibration. If so, the two states can become mixed and the '$d$-$d$' transition acquires extra intensity at the expense of the charge-transfer band. The mixing may be small but when the charge-transfer band is very intense, the augmentation of the '$d-d$' intensity in this way can be considerable.

## 4.5 'Two-Electron Jumps'

Sometimes, spin-allowed bands are much weaker than otherwise expected. There can be many reasons for this, most of which require more detailed analysis than we are able to present here. One particular case, however, can be discussed. It is well illustrated by the spectra of octahedral cobalt(II) species, an example being shown in Fig. 4-5. Three spin-allowed transitions are expected for these $d^7$ complexes, namely $^4T_{1g}(F) {\rightarrow} ^4T_{2g}$, ${\rightarrow} ^4A_{2g}$, ${\rightarrow} ^4T_{1g}(P)$ – see Chapter 3. The bands in Fig. 4-5 are so labelled. Note the weakness of the $^4T_{1g}{\rightarrow} ^4A_{2g}$ transition. The situation is quite typical of the spectra of octahedral cobalt(II) complexes. On occasion, the ${\rightarrow} ^4A_{2g}$ transition barely appears as a weak shoulder on the ${\rightarrow} ^4T_{1g}(P)$ band and can be missed. Why is this band so weak? We get the answer by looking at the $d^7$ correlation diagram, the spin-allowed part of which is shown in Fig. 4-6. Observe how the ground term $^4T_{1g}(F)$ in the weak field correlates with the strong-field configuration $t_{2g}^5 e_g^2$; $^4T_{2g}$ and $^4T_{1g}(P)$ with $t_{2g}^4 e_g^3$; and $^4A_{2g}$ with $t_{2g}^3 e_g^4$. At the strong-field limit, therefore, the transitions $^4T_{1g}(F){\rightarrow} ^4T_{2g}$, ${\rightarrow} ^4T_{1g}(P)$ involve the promotion of *one* electron from the $t_{2g}$ subset to the $e_g$ subset. On the other hand, the transition $^4T_{1g}(F){\rightarrow} ^4A_{2g}$ correlates with the promotion of *two* electrons from the $t_{2g}$ to the $e_g$ set. This is an example of a so-called 'two-electron jump'. It is intrinsically less probable than a one-electron jump and so the ${\rightarrow} ^4A_{2g}$ band is only weakly observed.

Of course, in real systems, the relative contributions of Coulomb and crystal-field effects are such as to place chromophores somewhere inbetween the weak- and strong-field limits. In that case, a real $^4T_{1g}(F) \rightarrow {}^4A_{2g}$ transition is not a pure two-electron jump, so that some intensity is observed.

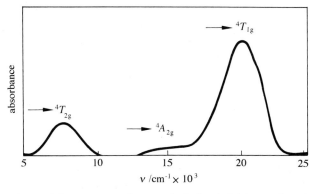

**Figure 4-5.** Spectrum of an octahedral cobalt(II) complex showing a weak $^4T_{1g} \rightarrow {}^4A_{2g}$ band.

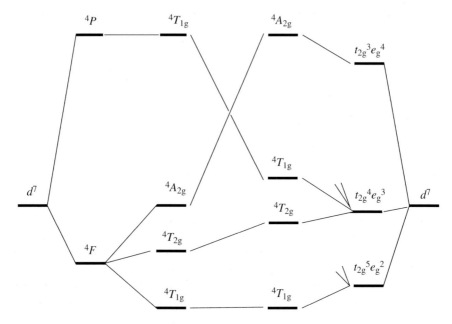

**Figure 4-6.** Partial correlation diagram for the spin-quartets of $d^7$ ions in octahedral symmetry.

---

**Box 4-3**

Absorption of *one* photon of light results in the relocation (with respect to space, spin or both) of *one* electron. It is possible, but extremely unlikely, that a second photon, together with its associated electronic rearrangement, can be absorbed before the ground state is re-acquired upon expulsion of a photon. It's unlikelyhood is because the lifetime of the excited state is typically only $10^{-18}$ seconds or so.

---

## 4.6 'Spin-Flip' Transitions

Here we comment on the *shape* of certain spin-forbidden bands. Though not strictly part of the intensity story being discussed in this chapter, an understanding of so-called spin-flip transitions depends upon a perusal of correlation diagrams as did our discussion of two-electron jumps. A typical example of a spin-flip transition is shown in Fig. 4-7. Unless totally obscured by a spin-allowed band, the spectra of octahedral nickel (II) complexes display a relatively sharp spike around 13,000 cm$^{-1}$. The spike corresponds to a spin-forbidden transition and, on comparing band areas, is not of unusual intensity for such a transition. It is so noticeable because it is so narrow – say 100 cm$^{-1}$ wide. It is broad compared with the 1–2 cm$^{-1}$ of free-ion line spectra but very narrow compared with the 2000–3000 cm$^{-1}$ of spin-allowed crystal-field bands.

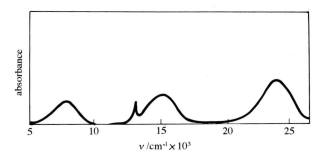

**Figure 4-7.** Spectrum of a typical, octahedral nickel(II) complex.

We briefly discussed the origin of the band widths of crystal-field spectra in Section 3.2. The broadening results from the way molecular vibrations affect ground and excited state energies differently. Sometimes, however, the response of ground and one or more excited states to bond length (and other vibrational) changes can be similar. The variations of all terms arising from the $d^8$ configuration with $\Delta_{oct}$ are shown in Fig. 4-8. This is another Tanabe-Sugano diagram of the type that we saw earlier for the $d^2$ configuration.

Notice how the energies of the $^1E_g$ and $^1T_{2g}$ terms from $^1D$ vary with $Dq$ in very nearly the same way as does that of the ground $^3A_{2g}$ term. Because of this parallelism, the transition energy from $^3A_{2g}(^3F) \rightarrow ^1E_g(^1D)$ hardly changes during the course of any vibration that affects the magnitude of $Dq$. The transition is thus seen as a

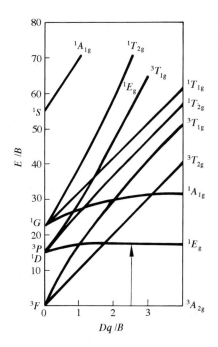

**Figure 4-8.** A term energy diagram for $d^8$.

sharp feature. Notice also that the same is not true of other spin-forbidden transitions. What is it about these particular spin-forbidden transitions that causes this parallel energy behaviour? A consideration of a complete $d^8$ correlation diagram helps provide the answer. The various electronic transitions we consider involve, in the strong-field limit, the rearrangement of electrons within or between the $t_{2g}$ and $e_g$ subsets. Some transitions, however, do not involve any spatial rearrangement but only a spin change. A transition from a spin-triplet term (with $S = 1$ and two unpaired electrons) to a spin-singlet term (with $S = 0$ and no unpaired electrons) can be achieved by reversing the sense of just one electron spin. If that is all that happens – that is, if the spatial distribution of the electrons remains unchanged – we refer to the transition as a 'spin-flip' transition. Since the spatial arrangements in ground and excited states for such a transition are the same, their responses to variations in the crystal-field strength (a space-only property) are the same, and the parallelism in diagrams like the one in Fig. 4-8 results. To emphasize this simple idea, we note that the transition within the $t_{2g}$ subset is an example of a 'spin-flip' transition. Note, once more, that not all spin-forbidden transitions involve only spin changes, so not all are of the spin-flip type and not all, therefore, are sharp. Our example of a 'spin-flip' transition is indicated by the arrow in Fig. 4-8.

## 4.7  The Effects of Temperature Change

We have just discussed one aspect of the shapes of '$d-d$' bands. For interest' sake, we finish this chapter with one more. It concerns the frequent, but not invariable, observation that band maxima may move somewhat towards the blue (higher frequency) end of the spectrum as a sample is cooled. In Fig. 4-9 are sketched potential energy curves for the ground and an excited electronic state. The 'ladders' represent the fundamental and various harmonic vibrational states associated with each electronic state. The vibrational states are typically separated by a (very) few hundred wavenumbers. Most molecules occupy the lowest vibrational state of the ground electronic state. One or more of the higher vibrational states are occupied to extents depending upon temperature and the Boltzmann distribution. Essentially no molecules occupy the excited electronic states. Electronic transitions promote molecules from members of the ground electronic state to members of the excited one. Those promoted from the higher lying vibrational states absorb light of a lower frequency than those promoted from the lowest vibrational state. As a sample is cooled, less molecules occupy the higher vibrational states and so the average electronic transition energy increases: the band maximum moves towards the blue, as sketched in Fig. 4-10. The progressive removal of the lower-energy transitions on cooling is referred to as the loss, or depletion, of 'hot bands'. The effect in Fig. 4-10 is not always observed for it depends upon the relative lateral displacements of the two potential wells in Fig. 4-9. Such variations are difficult, if not impossible, to calculate and hence predict.

In addition to these possible blue shifts, there is a general rule that the intensities of the '$d-d$' spectra of centrosymmetric molecules decrease with cooling while those of acentric chromophores do not. This follows from our discussions in Section 4.3 in which we noted that the origin of '$d-d$' intensities is vibronic (*i.e.* dynamically sourced) in centrosymmetric species but intrinsic (statically sourced) in non-centrosymmetric chromophores.

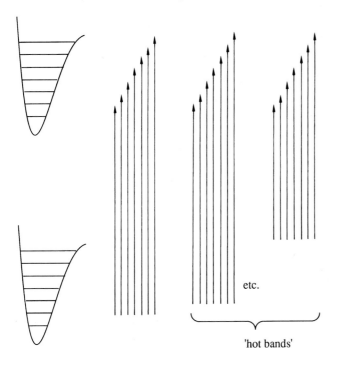

etc.

'hot bands'

**Figure 4-9.** Transitions occur from ground and vibrationally excited states of the ground electronic state to various vibrational components of the electronically excited state.

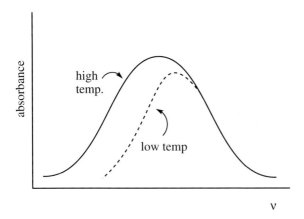

**Figure 4-10.** The qualitative appearance of the spectral band energies corresponding to the transitions in Fig. 4-9.

## 4.8  Summarizing Remarks

Crystal-field theory (and its successor, ligand-field theory, which we discuss in Chapter 6) forms a significant, and indeed large, part of standard texts and teaching courses in transition-metal chemistry for several interconnected reasons. It is remarkably successful at rationalizing a great body of spectroscopic and, as we shall see in the next chapter, magnetic data. It provides, at first level at least, a broadly accessible exercise in the exploitation of quantum mechanics and group theory. And, provided no questions of great detail or subtlety are put, it has predictive power, too. Again, we shall see more of that in the next two chapters. For the most part, its success and appeal are to be laid at the door of symmetry. The numbers of transitions to be expected in '$d-d$' spectra, and their patterns, are determined at root by symmetry in one guise or another. Matters with more chemical import inevitably involve factors of a continuously variable kind and when such considerations are incorporated into the crystal-field idea, predictions inevitably become more qualitative and arguments more subtle. A little of all this is evident in our descriptions of the 'violation' of electric-dipole selection rules. The rules are clear and sharp. Nature circumvents them a little by rendering them somewhat irrelevant. It all has been to do with the character of the wavefunctions in real systems, which are eigenfunctions of a, generally, complicated Hamiltonian. The labels we use are useful because they are approximately apt. However, because they are approximate, the rules get broken. How do we know these labels are reasonably apt? Because the rules are *only just* broken. The $d$ orbitals, though much affected by the molecular environment, tend to mind their own business.

## Suggestions for further reading

1. B.N. Figgis, *Introduction to Ligand Field,* Wiley, New York, **1966**, Chapter 9.
2. F.A. Cotton, *Chemical Applications of Group Theory*, 3rd. ed., Wiley, New York, **1990**.
3. P.W. Atkins, *Molecular Quantum Mechanics*, 2nd ed., Oxford University Press, Oxford **1983**.
4. P.W. Atkins, *Physical Chemistry*, 5th ed., Oxford University Press, Oxford, **1994**.
5. R.S. Berry, S.A. Rice, J. Ross, *Physical Chemistry*, Wiley, New York, **1980**.

   – All these references discuss selection rules from various points of view.

# 5 Spin and Magnetism

## 5.1 High-Spin and Low-Spin Configurations

In Chapter 3, we concentrated on the numbers and patterns of spin-allowed '$d$–$d$' transitions because, as we discussed in Chapter 4, they are usually more intense and obvious than the spin-forbidden ones. In fact, the perceived colours of most transition-metal complexes are determined by the spin-allowed '$d$–$d$' bands. Implicit throughout our discussions in Chapter 3 was an assumption that the spin-degeneracies of the crystal-field ground terms were the same as those of the corresponding free ions. We are not referring here to the fact that crystal-field terms arising from a free-ion term carry the same spin label, for that is always true. Rather, we are addressing the contest between interelectron repulsions and the crystal field. In the weak-field limit, the crystal-field ground term must be one of maximum spin-multiplicity because such is the case for the free ion itself and the free ion is the ultimate limit of a weak crystal field. To see if any different result is possible, we must move towards the strong field. We begin at the strong-field limit itself.

In Fig. 5-1 we represent possible strong-field ground configurations for $d^n$ ions in octahedral symmetry. Consider each $d^n$ ion in turn. For $d^1$ the lowest energy orbital arrangement (strong-field configuration) is that housing the solitary electron in the lower-lying $t_{2g}$ orbital subset. For $d^2$, it is similarly best to place both electrons within the $t_{2g}$ subset: the same goes for the three electrons of $d^3$. For $d^4$ we have a choice: we can place all four electrons within the lower-lying $t_{2g}$ set and suffer the crowding entailed in placing two electrons within the same orbital *or* we put three electrons in the $t_{2g}$ set and one in the $e_g$ set and suffer instead the promotion energy $t_{2g} \rightarrow e_g$; $\Delta_{oct}$. The choice in any particular case depends upon the relative energy penalties incurred from interelectron repulsion or from the crystal field. In the context of the present discussion, the penalty from interelectron repulsion is often called the *pairing energy*, $P$. Now, notice also, that all four electrons will take parallel spins in the configuration $t_{2g}^3 e_g^1$ because of Hund's rule while two electrons must pair up their spins in $t_{2g}^4$. These two arrangements or configurations are called high-spin or low-spin, as appropriate. We refer here to the net spin of an electronic arrangement, that is, to how many unpaired electrons there are. One may formalize the result with the inequalities given in Eq. (5.1).

$$\text{for high spin:} \quad \Delta_{oct} < P$$

$$\text{for low spin:} \quad \Delta_{oct} > P$$

(5.1)

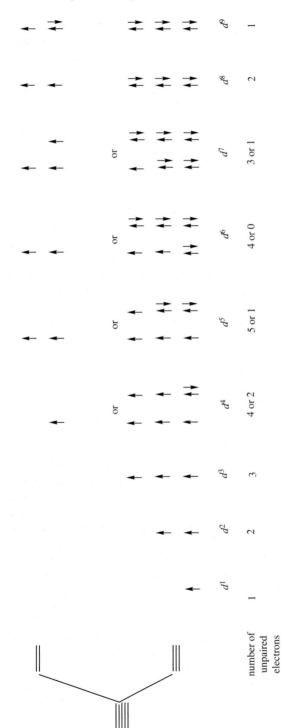

**Figure 5-1.** High- and low-spin arrangements of *d* electrons in strong octahedral fields.

For the $d^5$ configuration, we similarly have two choices corresponding to the strong-field configurations $t_{2g}^5$ and $t_{2g}^3 e_g^2$ for low- and high-spin arrangements respectively. The reader may ask why we don't consider an intermediate-spin arrangement $t_{2g}^4 e_g^1$. There is an 'all-or-nothing' reason. If $\Delta_{oct}$ is too great to allow the promotion of one electron from $t_{2g}^5$ it is still too great to allow the double promotion to $t_{2g}^3 e_g^2$. Conversely, if the pairing energy is too great to allow the forming of one pair of electrons, it is too great to yield two pairs. No octahedral $d^5$ ions are known with an intermediate-spin arrangement. The same is not true for molecules with other symmetries, $e.g.$ planar, but that is not at issue here.

The rest of Fig. 5-1 is completed in like manner. The $d^n$ ions with $n = 4,5,6,7$ offer two choices – high- or low-spin – while for $n = 1,2,3,8,9$ only one arrangement of lowest energy is possible. It is left as an exercise for the reader to construct a similar diagram for $d^n$ ions in a tetrahedral field and show that high- and low-spin choices exist for $n = 3,4,5,6$ but not for $n = 1,2,7,8,9$. Few if any examples of low-spin tetrahedral $d^n$ complexes exist, however, because $\Delta_{tet}$, being only $4/9\Delta_{oct}$, is not usually sufficient to prevent the $e - t_2$ promotion $i.e.$ $P > \Delta_{tet}$ always.

For the octahedral case in Fig. 5-1, we include mention of the number of unpaired electrons associated with each arrangement. For real molecules we could use this to determine which configuration is lowest in energy – whether $\Delta_{oct}$ or $P$ were the greater – if only we had some experimental method of measuring the number of unpaired electrons. There is such a method and it depends upon the interaction of these molecules with a magnetic field.

# 5.2  The Qualitative Origin of Paramagnetism

*All* substances interact with a magnetic field – there are no exceptions. Substances may be subdivided according to their manner of interaction with magnetic fields in various ways. An old classification which is directly empirical and generally useful is to group materials together which are either a) repelled, b) attracted or c) attracted very strongly. Only the latter are commonly recognized in everyday life. They are called *ferromagnets*. The group is not large – soft iron, cobalt, a few other metals and alloys, as well as a small number of special compounds. Their interaction with magnetic fields is many orders of magnitude stronger than that of the materials in groups a) or b). The origins of their ferromagnetic property lies in cooperative interactions between molecules or atoms and their study properly lies within the realm of physics. Fascinating though they are, we have no more to say about them.

Substances in group a) are repelled by a magnetic field and are called *diamagnets*. Diamagnetism is a *universal* atomic (and, hence, molecular) property and is generally very small in magnitude. That a bar magnet actually does repel a piece of paper, for example, can only be demonstrated with rather delicate apparatus. The much smaller, though nevertheless extensive, group of materials that are weakly attracted by a magnetic field define *paramagnets*. Many transition-metal compounds fall into this class. Paramagnetism, when it is present, is generally larger (often much larger) than diamagnetism. So although diamagnetism is universal in atoms – and hence molecules – it is almost invariably swamped by the paramagnetic effect when that is present. Even so, the attraction of copper sulphate crystals, which are paramagnetic, to a magnet can only be observed, once more, with delicate apparatus. Diamagnetism is the essence of chemical shifts in nuclear magnetic resonance spectroscopy, for example, and so is a very important topic elsewhere in chemistry. For our present area, however, diamagnetic effects are treated as corrections to any paramagnetism. Our remarks on magnetism from now on are therefore confined exclusively to the case of paramagnetism.

It is convenient to begin with a classical picture. Paramagnetic substances are considered to comprise molecules with permanent magnetic dipole moments, $m$. They may be regarded here as miniature bar magnets. *All* discussions of bulk magnetism concern molecules *en masse* – molecular ensembles. In the absence of an applied field, a paramagnetic sample comprises molecules whose permanent magnetic dipoles are oriented randomly, as indicated in Fig. 5-2a, because of the ever present thermal agitation to which all molecular ensembles are subject. On application of an external magnetic field, these molecular magnets will *tend* to align parallel to the field. We can safely neglect here any tendency of the molecular

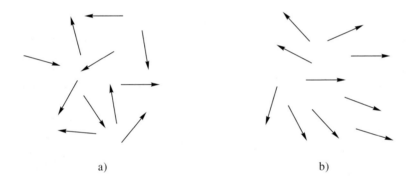

a)                                              b)

**Figure 5-2.** In the absence of an applied magnetic field a), the molecular magnetic dipoles are randomly oriented; on application of an external field b), the dipoles *tend* to orientate parallel to the field.

magnets to align parallel because of their magnetic interaction with each other: that is because the strength of the magnetic field from any one such magnet is tiny as compared with the externally applied field (which could be thought of as arising from $10^{30}$ or more such aligned molecular magnets). Once again, thermal agitation will prevent their aligning perfectly (Fig. 5-2b). Any particular molecular magnet aligned exactly parallel to the field, $B$, will possess energy $-m \cdot B$, while the energy of any such magnet aligned exactly antiparallel will be $+m \cdot B$ relative to their energy before the application of the field. Molecular magnets oriented between these extremes will acquire energies between these limits. In Fig. 5-3, we represent the thermal distribution of an ensemble of molecular magnets in an applied field. Since, on average, the molecular magnets tend to align more with the field than against it, the

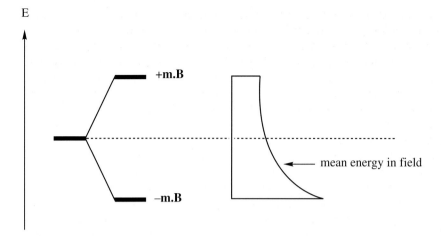

**Figure 5.3.** The classical picture: the energies of dipoles varies continuously from parallel alignment with the applied magnetic field ($-m \cdot B$) to antiparallel ($+m \cdot B$). On the right is shown the distribution of molecules that results and the lower mean energy of the ensemble relative to the field-free environment.

average energy of the ensemble is less in the presence of an external field than in its absence, as also indicated in the figure. The phenomenon of paramagnetism thus arises because the energy of an ensemble of molecular magnets decreases – the ensemble acquires more stability – on application of a magnetic field: the sample is attracted by the field. It is attracted because, if otherwise unconstrained, the sample will move from a place of no field into a field for, by doing so, its energy is reduced. The energy change involves a *redistribution* between that for molecular alignment and that for thermal agitation; there is no exchange of energy with the magnetic field itself. Any (very, very slight) warming of the sample on application of the field is soon dissipated, an effect whose reverse usage allows for the technique of adiabatic cooling (see Box 5-1)

---

**Box 5-1**

The technique of adiabatic cooling is used to achieve temperatures lower than can be obtained by the conventional techniques of immersing a sample in liquid helium under low pressure, a process which might cool a sample to around 1.6 K. To cool a sample further, one can proceed as follows. The method depends upon the sample being held within a paramagnetic container. During the conventional cooling process, the paramagnetic holder is held within a strong magnetic field. According to the usual Boltzmann statistics, more molecules of the container occupy the lower energy levels than the higher ones. When equilibrium has been achieved, the magnetic field is switched off. The split energy levels return to a degenerate condition and the distribution of the molecules of the paramagnetic container reverts to a situation like that on the extreme left of Fig. 5-3. This change in distribution requires an input of thermal energy. That energy is taken from the environment, including the sample. Cooling of samples down to millikelvin levels can be achieved in this way.

The greater the magnitude of the applied field, the greater the energy difference between parallel and antiparallel alignment of the molecular magnets and the less able is the thermal agitation to randomize the molecular orientations. So, as shown on the right in Fig. 5-3, the mean energy of the ensemble decreases with increasing magnetic field strength. This effect is exploited in various 'force technique' methods of measuring magnetism. The Gouy method is illustrated in Fig. 5-4. The sample is taken in the form of a cylindrical rod – or, in the case of powders or solutions, contained within a cylindrical glass tube – to make integrations trivial. The sample is suspended from a chemical balance, or any other force-measuring device, and placed so that its bottom end lies near the strongest part of a magnetic field and its

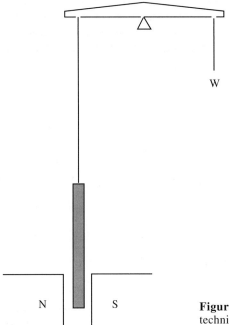

W

N          S

**Figure 5-4.** Schematic arrangement for the Gouy technique. The sample must be placed asymmetrically (vertically) in the magnetic field.

top end near the weakest part. The source of the magnetic field is conveniently provided by an electromagnet (or permanent magnet). Suppose the sample is balanced in the absence of the magnetic field. On switching on the electromagnet, (or on introducing the permanent magnet) the sample will move toward the stronger part of the field because more of the sample will enter a strong field and the energy loss, as in Figure 5-3, will be greater. On re-taring the balance, we observe the sample to weigh more in the field than out of it. In SI units, the force, $F$, on the cylindrical sample of cross-sectional area, $a$, is

$$F = \frac{1}{2}\mu_0 a \bar{\chi} H^2 \qquad (5.2)$$

where $H$ is the strength of the magnetic field at the bottom of the sample and $\overline{\chi}$ is the mean volume susceptibility of the sample (see Box 5-3 for units).

We now modify the classical picture set out above to accommodate quantum mechanics.[*] Instead of talking about permanent molecular magnetic dipoles, we postulate (for the moment, but we shall return to the point shortly) that paramagnetic molecules are those whose occupied energy levels (usually the ground or near-ground levels) possess a degree of degeneracy that is removed by the application of a magnetic field. (We discuss an exception to this under 'temperature independent paramagnetism', later). Instead of Fig. 5-3, we construct Fig. 5-5. The splitting of the levels (a) in a magnetic field is symmetrical (b), as it was in the classical picture. The distribution of molecules amongst the component states is no longer equal in the presence of the applied field. An appropriate (Boltzmann) population histogram for the distribution of molecules amongst these states is shown in (c). Once again, we see that the mean energy of the molecular ensemble is less after application of the field than before it and the phenomenon of paramagnetism follows once more.

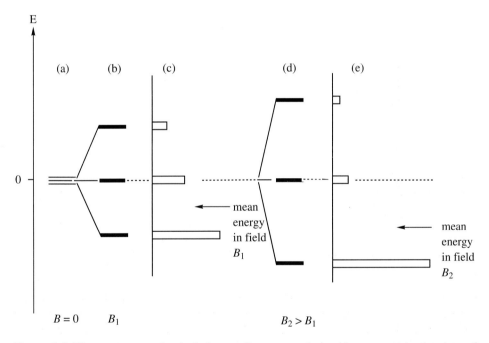

**Figure 5-5.** The quantum mechanical picture: discrete population histograms take the place of continuous distributions. The overall paramagnetism increases with increasing field strength.

---

[*] A *careful* classical analysis of *all* magnetization phenomena shows them to vanish identically! The interactions of matter with magnetic fields that we observe on a day-to-day basis are purely quantum phenomena just as the existence of magnetic fields is a relativistic phenomenon. The classical prediction of vanishing magnetism is really as great a failure of the classical regime as the better-known 'Ultraviolet catastrophe'.

Also illustrated in Fig. 5-5, is the greater splitting (d) that follows an increased magnetic-field strength, whose splitting is, in fact, linear in the field strength, *B*. The Boltzmann distribution changes accordingly to (e), the mean energy decreases, and the force on the sample in a Gouy experiment increases. Quantitative analysis shows that the force on the sample varies linearly with the field strength until that field grows very large when the effect falls off and one observes (theoretically and experimentally) the phenomenon of *magnetic saturation*, as shown in Fig. 5-6. Under normal laboratory conditions, magnetic saturation is rarely observed.

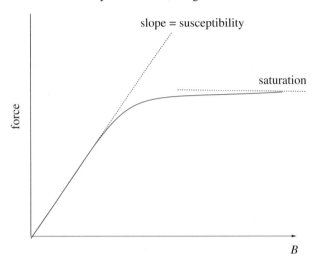

**Figure 5-6.** Saturation and the definition of magnetic susceptibility, $\chi$.

---

**Box 5-2**

Classically, 'saturation' occurs when the field is so strong and/or the thermal agitation (temperature) is so feeble that all the molecular dipoles are aligned with the field. Then, of course, increasing the applied field more is unable to cause any further alignment.

---

Rather than quote some (mass normalized) force on the sample at each of several field strengths, it is sufficient to report the slope of the linear part of the curve in Fig. 5-6. This slope is called the *magnetic susceptibility* of the sample. Units for susceptibility, $\chi$ , and related quantities to be discussed in this section are reviewed in Box 5-3.

The effect of temperature upon the situation in Fig. 5-5 is to modify the Boltzmann distribution. Lowering the temperature depopulates the higher-lying energy levels in favour of the lower. Therefore, susceptibility increases with decreasing temperature. Quantitative studies of the simple (first-order[*])

---

[*]'First-order' means that we consider nothing beyond that described here. In 'second-order', we would include the effects of mixing between ground and excited states brought about by the magnetic field. This is briefly discussed under 'second-order Zeeman effects' later.

---

**Box 5-3**

*Units for magnetochemical quantities*

$$B = H + 4\pi M \qquad : \qquad CGS$$

$B$ and $H$ can be quoted in gauss

$$B = \mu_o(H + M) \qquad : \qquad SI$$

$B$ and $H$ are quoted in tesla (T): $1T = 10^4$ gauss

Several texts describe $B$ as the applied magnetic field and $H$ as the field in the sample. This is incorrect since both $B$ and $H$ exist inside and outside the sample. $B$ is the magnetic field associated with a current loop source. $H$ is the magnetic field associated with a (fictitious) magnetic point monopole: see *Magnetism and Ligand-Field Analysis* by M. Gerloch (Cambridge University Press, 1983). $B$ and $H$ differ very little with respect to magnitude or direction for weakly magnetic (non-ferromagnetic) materials. Many confusions about $B$ and $H$ arise because of this.

$$B/H = \mu_o(1 + \chi) \qquad : \qquad SI$$
$$B/H = 1 + 4\pi\chi \qquad : \qquad CGS$$

where $\chi$ is the volume susceptibility (dimensionless). Gram susceptibility, $\chi_g$, is defined by

$$\chi_g = \chi/\rho$$

where $\rho$ is the density of the sample. Molar susceptibility, $\chi_M$, is defined by

$$\chi_M = \chi_g M$$

where $M$ is the molecular weight of the material. In the SI system, $\chi_M$ is measured in $m^3 mol^{-1}$; in the CGS system it is measured in $cm^3 mol^{-1}$. To convert $\chi_M$ values quoted in the CGS system into SI values, multiply by $4\pi \times 10^{-6}$.

Effective magnetic moments, $\mu_{eff}$, defined in (5.4), are quoted in Bohr magnetons in *either* SI or CGS systems. In the CGS system the Bohr magneton is $0.92731 \times 10^{-20}$ erg gauss$^{-1}$ whilst in the SI system it is $0.92731 \times 10^{-23}$ A m$^2$ molecule$^{-1}$. The magnetic moment, $\mu_{eff}$, is then $2.8279(\chi_M T)^{1/2}$ B.M. in the CGS system, and $7.9774 \times 10^2(\chi_M T)^{1/2}$ B.M. in SI. These expressions yield the same numerical values for $\mu_{eff}$, so that expressions like (5.6), (5.7), (5.10) and (5.11) *etc.* remain valid in both CGS and SI systems.

---

circumstances of Fig. 5-5 predict that paramagnetic susceptibilities are inversely proportional to temperature. In the later years of the last century, Pierre Curie summarized a wealth of experimentation on paramagnetic substances with the law that bears his name:

$$\text{Curie's Law} \qquad \chi = C/T \qquad\qquad (5.3)$$

where $C$ is called Curie's constant. As a plot of $\chi$ versus $1/T$ is a straight line, according to this law and first-order theory, there is no need to report $\chi$ values at different temperatures (see Fig. 5-7). It is sufficient to report the slope of such a

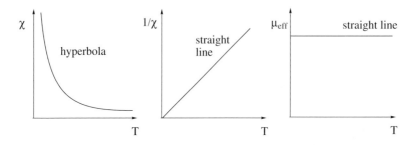

**Figure 5-7.** Curie Law behaviour of $\chi$, $1/\chi$ and $\mu_{\text{eff}}$ with respect to temperature.

relationship, and that slope is just $C$. For historical reasons we need not pursue, it is conventional instead to report a quantity known as the *effective magnetic moment*, $\mu_{\text{eff}}$, for a sample where $\mu_{\text{eff}}$ is proportional to the square root of $C$:

$$\mu_{\text{eff}} = \frac{e\hbar}{mc^2}\sqrt{\chi T} \qquad (5.4)$$

which is approximately the same as the relationship in Eq. (5.5).

$$\mu_{\text{eff}} = 2.828\sqrt{\chi T} \qquad (5.5)$$

Insofar that Curie's law is true, $\mu_{\text{eff}}$ is independent of temperature, for that is how we arrived at Eq. (5.5). In practice, Curie's law is rarely obeyed exactly and, occasionally, it is quite seriously flouted. Nevertheless it is still conventional to quote $\mu_{\text{eff}}$ values but it is then necessary to quote them over a range of temperatures. Although we might just as well report susceptibility values in these circumstances, conventions die hard. In any case, the temperature variation for $\mu_{\text{eff}}$ immediately and transparently reveals any departures from Curie's Law in a way that the temperature variation of susceptibilities might not.

We have seen how the phenomenon of paramagnetism follows from the assumption that paramagnetic molecules possess appropriate degenerate states which split in a magnetic field. The question arises as to 'what molecular property leads to these degeneracies?' and hence 'what does a magnetic susceptibility ultimately measure?'. The answer is that paramagnetic moments or susceptibilities are a measure of angular momentum – both spin- and orbital-angular momentum or the total angular momentum when that is a more appropriate quantity (*i.e.* when the effects of spin-orbit coupling are large). Physically, and in outline only, it is simple to see why this should be so. All measurements involve an interaction between the system and the apparatus. Interactions only take place between quantities of the same kind. Orbiting or spinning electrons generate magnetic fields; applied magnetic fields are generated by circulating electrons (current loops).

The angular momenta of atoms are described by the quantum numbers $L$, $S$ or $J$. When spin-orbit coupling is important, it is the total angular momentum $J$ which is a constant of the system. A group of atomic wavefunctions with a common $J$ value – akin to a term, as described in Section 3.6 – comprise $(2J + 1)$ members with $M_J$

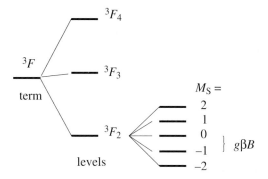

**Figure 5-8.** Spin-orbit coupling splits the $^3F$ term into three levels $^3F_J$. An externally applied magnetic field splits up the levels into their $M_J$ components.

values ranging $J$, $J–1$,...$–J$. The $(2J + 1)$ degeneracy of such a *level,* as it is called, is removed completely by an external magnetic field, as shown in Fig. 5-8. The result, as we have seen, is paramagnetism. Quantitative analysis of these circumstances yields the formula in Eq. (5.6)

$$\mu_{\text{eff}} = g\sqrt{J(J+1)} \tag{5.6}$$

where g is defined as in Eq. (5-7).

$$g = 1 + \frac{J(J+1)+S(S+1)-L(L+1)}{2J(J+1)} \tag{5.7}$$

The $L$ and $S$ values are those from which the $J$ value was formed via the vector coupling rule.[*] These formulae strictly apply only for the magnetism of free-ion levels. They provide a good aproximation for the magnetism of lanthanide complexes, as we shall note in Chapter 10, but provide no useful account of the magnetic properties of $d$ block compounds.

A corresponding formula (Eq. 5.8), due to Van Vleck, has been derived for free atoms in which the effects of spin-orbit coupling can be ignored.

$$\mu_{\text{eff}} = \sqrt{L(L+1)+4S(S+1)} \tag{5.8}$$

Again, however, this is strictly applicable only for free ions. Even though spin-orbit coupling is much less important for the first row of the $d$ block, this formula provides a far less good approximation for $d$-block complexes than Eq. (5.6) does for lanthanide complexes. The reason is that the ground, and other, terms in these $d$ complexes differ grossly from those of the corresponding free ion. These differences are one result of the crystal field.

---

[*](see 'Orbitals, Terms and States', Ch. 4).

## 5.3 'Orbital Quenching' and the 'Spin-Only' Formula

We have nearly made contact again with our crystal-field discussions in Section 5.1. In order that magnetic measurements be useful to us, however, we need to study the angular momenta associated with various crystal-field terms. First, we recall that no crystal field affects the spin angular momentum of any given free-ion term.[*] As we observed in Section 3.7, crystal fields act only upon the spatial parts of wavefunctions. If a parent free-ion term has a spin-degeneracy $(2S + 1)$, so also do the crystal-field terms that derive from it. Thus, any changes in angular momentum brought about by a crystal field concern only the orbital (spatial) part.

Consider the orbital angular momentum of a free-ion $^3F$ term. Here $L = 3$ and the orbital degeneracy is 7. Application of Van Vleck's formula (5.8) predicts an effective magnetic moment,

$$\mu_{\text{eff}} = \sqrt{3 \cdot 4 + 4 \cdot 1 \cdot 2} = \sqrt{20} \tag{5.9}$$

In octahedral symmetry, the $F$ term splits into $A_{2g} + T_{2g} + T_{1g}$ crystal-field terms. Suppose we take the case for an octahedral nickel(II) complex. The ground term is $^3A_{2g}$. The total degeneracy of this term is 3 from the spin-multiplicity. Since an $A$ term is orbitally (spatially) non-degenerate, we can assign a fictitious $L_{\text{eff}}$ value for this of 0 because $2L_{\text{eff}} + 1 = 1$. We might employ Van Vleck's formula now in the form

$$\mu_{\text{eff}} = \sqrt{L_{\text{eff}}(L_{\text{eff}} + 1) + 4S(S + 1)} \tag{5.10}$$

---

**Box 5-4**

Strictly, $L$ is defined only as a quantum number for a spherical environment - the free ion. The use of $L_{\text{eff}} = 0$ for $A$ terms or $L_{\text{eff}} = 1$ for $T$ terms on the grounds that $(2L_{\text{eff}} + 1)$ equals the degeneracy of these terms is, however, legitimate as used here. There is a close parallel between the quantum mechanics of $T$ terms in octahedral or tetrahedral symmetry on the one hand, and of $P$ terms in spherical symmetry on the other.

---

and so predict a magnetic moment, $\mu_{\text{eff}} = \sqrt{8}$. In other words, for this $^3A_{2g}$ ground term, the crystal field has completely removed the orbital angular momentum. We say that the orbital angular momentum has been *quenched*.

Now take the case for an octahedral vanadium(III) ion. For $d^2$, the ground term is $^3T_{1g}$. The spatial degeneracy of a $T$ term is three-fold and we describe this with $L_{\text{eff}} = 1$. Using (5.10) we find $\mu_{\text{eff}} = \sqrt{10}$. So for this $^3T_{1g}$ term, the crystal field has quenched some, but not all, of the angular momentum of the parent free ion $F$ term.

---

[*] Except, apparently, insofar that for some $d^n$ configurations a strong crystal field may bring about a 'low-spin' configuration as described in Section 5.1. However, in these cases, the corresponding ligand-field terms correlate with excited free-ion terms which still have the same spin-multiplicity as that of the strong-field term.

Analogous arguments apply to the various ground terms of octahedral or tetrahedral $d^2$, $d^3$, $d^7$ and $d^8$ complexes.

The ground term for octahedral or tetrahedral $d^5$ complexes is orbitally non-degenerate ($^6A_{1g}$ or $^6A_1$ respectively) and so, once again, we expect no orbital contribution to the magnetic moment. The situations for $d^1$, $d^4$, $d^6$ and $d^9$ are special in one respect. When the ground term is $T_{2g}$ or $T_2$, we have $L_{eff} = 1$ and partial quenching as before. When, however, it is an $E_g$ or $E$ term, our simple ploy of using $L_{eff}$ doesn't work and, in fact, these terms give rise to no orbital contribution at all (see Section 5.5). They are known as *non-magnetic doublets* because of this. Beware of this jargon, by the way, for the lack of magnetism only refers to the *orbital* contribution; magnetism still arises from the spin angular momentum of $^2E_{(g)}$ or $^5E_{(g)}$ terms. Some explanation for the lack of orbital angular momentum for these cubic-field $E$ terms will be given shortly. Meanwhile, we note that the phenomenon applies only to these $E$ terms arising in strict octahedral or tetrahedral symmetry. Lower symmetry environments also define $E$ terms on occasion, but these are not generally bereft of orbital angular momentum.

Altogether then, Van Vleck's formula for free ions is inappropriate for the paramagetism of crystal-field terms. Crystal fields partly or completely quench the orbital angular momentum (spatial degeneracy). To the extent that such quenching is complete, we might consider using the limiting case of Van Vleck's equation where the magnetism is ascribed to the spin-angular momentum alone. This yields the so-called *spin-only formula* (Eq. 5.11).

$$\mu_{so} = \sqrt{4S(S+1)} \tag{5.11}$$

Now the total spin-angular momentum quantum number $S$ is given by the number, $n$, of unpaired electrons times the spin angular momentum quantum number $s$ for the electron, that is, $S = n/2$. Substitution of this relationship into Eq. (5.11) yields an alternative form of the spin-only formula,

$$\mu_{so} = \sqrt{n(n+2)} \tag{5.12}$$

which directly expresses the effective magnetic moment in terms of the number of unpaired electrons. Of course, as discussed above, there may be some orbital contribution in any particular complex, but to the extent that the spin-only formula is appropriate, it provides the measure of the number of unpaired electrons that we required at the end of Section 5.1.

---

**Box 5-5**

| number of unpaired electrons | 1 | 2 | 3 | 4 | 5 |
|---|---|---|---|---|---|
| $\mu_{so}$/BM | 1.73 | 2.83 | 3.87 | 4.90 | 5.92 |

---

By way of example, a $d^5$ iron(III) complex with a magnetic moment close to 1.73 must, by reference to Fig. 5-1, be low-spin with $\Delta_{oct} > P$ since an iron(III) complex with $\Delta_{oct} < P$ would have a magnetic moment of 5.92.

## 5.4 Orbital Contributions

We have seen that there are orbital contributions to the magnetic moments of complex ions with $T$ ground terms but not with $A$ or $E$ terms. These rules are only approximate, for the quantitative theory of paramagnetism is a rather complicated affair that we can do no more than skim in this book. One group of refinements to our earlier statements recognizes the over simplistic description of the true molecular wavefunctions. We saw the same sort of thing in our discussion of electric-dipole selection rules in Section 4.3. Although interelectron repulsion and crystal-field energies are much greater than spin-orbit coupling energies, the latter cannot be ignored when we look at properties as sensitive as '$d-d$' intensities or magnetic susceptibilities. Let us, therefore, take a second look at the ground wavefunctions of octahedral $d^8$ ions.

Ignoring spin-orbit coupling, it is exact to label these ground wavefunctions with the term label $^3A_{2g}$, and it is equally exact to use the spin-only formula for the magnetic moment. Strictly, however, we should not ignore spin-orbit coupling for it causes some mixing between the $^3A_{2g}$ ground term and, for example, the higher-lying $^3T_{2g}$ term (Eq. 5.13).

$$\psi(`^3A_{2g}`) = \psi(^3A_{2g}) + c \; \psi(^3T_{2g}) \tag{5.13}$$

The mixing has nothing to do with the possibility of any molecules populating the $^3T_{2g}$ term, which is typically 12,000 cm$^{-1}$ above the ground state term. The population of such a term is of the order $e^{-12000/200}$ at room temperature ($kT \approx 200$ cm$^{-1}$ at 300 K), which is absolutely negligible. The mixing arises because a description of the molecular Hamiltonian in terms of Eq. (5.14) is incomplete and should be replaced with Eq. (5.15).

$$\mathcal{H}_1 = \sum_i^n \mathcal{H}_{H-like}(i) + \sum_{i<j}^n \frac{e^2}{r_{ij}} + V_{CF} \tag{5.14}$$

$$\mathcal{H}_2 = \sum_i^n \mathcal{H}_{H-like}(i) + \sum_{i<j}^n \frac{e^2}{r_{ij}} + V_{CF} + \lambda L.S \tag{5.15}$$

Wavefunctions like $\psi(^3A_{2g})$ are eigenfunctions of $\mathcal{H}_1$ but those of $\mathcal{H}_2$ are slightly different. In writing them like $\psi(`^3A_{2g}`)$ we merely indicate what the wave functions are most nearly like. The functions $\psi(`^3A_{2g}`)$ can be expressed in literally an infinite variety of ways, although these must be explicitly determined. One way which is rather convenient for further discussion is that shown in Eq. (5.13). In words, we might say that 'under spin-orbit coupling, the '$^3A_{2g}$' term looks as if it contains some $^3T_{2g}$ character'.

The extent of the mixing – the magnitude of $c$ in (5.13) – is proportional to the cause of the mixing and inversely proportional to the energy separation of the original terms being mixed, that is, $c \propto \lambda/\Delta_{oct}$ in this case. The admixture of $T$ character into the formal $A$ ground term implies admixture of orbital angular momentum. Detailed theory yields the expression,

$$\mu_{eff} = \mu_{so}(1 - 4\lambda/\Delta_{oct}) \tag{5.16}$$

as the result of this on the paramagnetic moment. For an octahedral nickel(II) complex, for example, $\Delta_{oct}$ is typically 12,000 cm$^{-1}$, $\lambda$ is $-315$ cm$^{-1}$ (negative because $d^8$ is a more-than-half-filled shell), $\mu_{so} = 2.83$ and so, from Eq. (5.16), we calculate $\mu_{eff} = 3.13$. This is, indeed, the sort of value that is typically observed for such complexes. For octahedral chromium(III) complexes, $d^3$, with formal $^4A_{2g}$ ground terms, the less-than-half-filled $d$ shell means $\lambda$ is positive – though rather less than that for $d^8$ in magnitude – and Eq. (5.16) predicts magnetic moments somewhat *less* than that given by the spin-only formula, and *that* is also observed in practice. The formula in Eq. (5.16) applies to all cubic-field systems having an $A_{2g}$ ground term.

---

**Box 5-6**

Reminder: The one-electron spin-orbit coupling coefficient, $\zeta$, is intrinsically positive. The many-electron spin-orbit parameter $\lambda$ is defined by

$$\lambda = \pm \; \zeta/2S$$

and $\lambda$ takes positive values for a less-than-half-filled shell, and negative values for a more-than-half-filled shell. These signs conform with Hund's third rule that minimum $J$ values lie lowest in energy for less-than-half-filled shells, and highest in energy for more-than-half-filled ones.

---

**Box 5-7**

Some useful spin-orbit coupling coefficients:

| Ion | Ti$^{3+}$ | V$^{3+}$ | Cr$^{3+}$ | Mn$^{3+}$ | Fe$^{2+}$ | Co$^{2+}$ | Ni$^{2+}$ | Cu$^{2+}$ |
|---|---|---|---|---|---|---|---|---|
| $\zeta$/cm$^{-1}$ | 55 | 210 | 270 | 350 | 410 | 530 | 630 | 830 |
| $\lambda$/cm$^{-1}$ | 155 | 105 | 90 | 88 | $-102$ | $-177$ | $-315$ | $-830$ |

---

A similar expression has been derived for cubic-field complex ions having an $E$ ground term (Eq. 5.17).

$$\mu_{eff} = \mu_{so}(1 - 2\lambda/\Delta_{oct}) \tag{5.17}$$

The basis for this formula is just the same as described above but, in this case, spin-orbit coupling admixes the higher-lying $T_{2(g)}$ term wavefunctions into the ground $E_{(g)}$. The coefficient 2 in Eq. (5.17) rather than the 4 in Eq. (5.16) arises from the different natures of the wavefunctions being mixed together.

These two formulae describe orbital contributions to ground $A$ or $E$ terms that arise by so-called second-order spin-orbit coupling with appropriate excited wavefunctions. For complexes with $T$ ground terms, the effects of spin-orbit coupling are first-order and rather complicated. We describe them in barest outline only. Recall how a free-ion $^3F$ term splits under spin-orbit coupling to give three so-called *levels*, $^3F_4$, $^3F_3$, $^3F_2$, with total angular momenta $J$ values of 4, 3 and 2 respectively. In a similar way, a $^3T_{1g}$ crystal-field term, for example, splits into three components with $J$ values of 2, 1 and 0 (we use the vector coupling rule with $S = 1$ and $L_{eff} = 1$). The energy separations between these components are typically of the order $50-500$ cm$^{-1}$ for first row transition-metal complexes and so there are significant molecular populations of each of them. Each component generally splits up in an applied magnetic field and gives rise to a contribution to the magnetic moment which depends, in part, upon the population of that component. This immediately tells us that the magnetic moments of such systems are generally temperature dependent, at least because the populations of the original components are temperature dependent.

In addition to this complex behaviour is the *second-order Zeeman effect*. The splitting of a degenerate set of wavefunctions by an applied magnetic field, as illustrated in Fig. 5-5, is called a *first-order Zeeman splitting*. Consideration of this effect alone always yields a Curie-Law behaviour and temperature-independent magnetic moments. However, spin-orbit coupling is not the only mechanism that can scramble wavefunctions. The applied magnetic field itself effectively polarizes the wavefunctions as well as changing their energies in such a way that any one wavefunction looks as if it has some of the character of (most) other wavefunctions admixed. This is the second-order Zeeman effect. (In truth, there is only *one* Zeeman effect and these names merely refer to two parts of a mathematical expansion). The amount of such magnetic-field-induced mixing is (again) proportional to the cause – the magnetic moment operator – and inversely proportional to the energy separations of the wavefunctions being admixed. For typical laboratory magnetic fields, the numerators here are of the order $0.1-1$ cm$^{-1}$ so the second-order Zeeman effects are often very small indeed. However, in the case of an ion with a formal orbital triplet ground term split into fairly close-lying spin-orbit components, these effects can be quite important. This is because the mixing induced by the magnetic field, though proportional to the small magnetic-field term (small numerator), is inversely proportional here to a fairly small energy denominator. They *can* be calculated, but a discussion of how is outside our present scope. Suffice it to say that the magnetic moments of ions with formal orbital triplet ground terms have orbital contributions which are not simply estimable by the tactic of using $L_{eff} = 1$ as we described earlier. Furthermore, they tend to vary considerably with temperature.

For ions with formal orbital singlet ground terms, it is often quite adequate to ignore second-order Zeeman terms since any magnetic-field-admixed wavefunctions are energetically well removed from the ground state. There is one type of situation, however, when these small effects are observable because they are the *only* contribution to magnetic susceptibilities. The classic case is that of low-spin octahedral cobalt(III) complexes. From Fig. 5-1, we note the ground strong-field configuration to be $t_{2g}^6$. The subshell is full and hence uniquely defined. There is no spin- or space-degeneracy associated with this electronic arrangement and it is labelled $^1A_{1g}$. Having no degeneracy at all, it cannot be split by an applied magnetic field and so we expect no paramagnetism, just the diamagnetism that is a property of all atoms. Experimentally, however, such systems are observed to be slightly paramagnetic with magnetic moments of about 0.5 BM that vary with temperature according to a square-root relationship, $\mu_{\text{eff}} \propto \sqrt{T}$. Actually, this temperature dependence of $\mu_{\text{eff}}$ is misleading because of the way in which $\mu_{\text{eff}}$ is defined in Eq. (5.4). It conceals the more interesting fact that the *susceptibility* is independent of temperature. Indeed, the phenomenon we describe here is called *temperature-independent paramagnetism (T.I.P.)* Its origin is as follows.

The ground state is indeed non-magnetic as discussed above, but only in first order. That is, there is no first-order splitting – no first-order Zeeman effect. There is however, a second-order Zeeman effect in which the applied magnetic field mixes various excited-state character into the ground state. Some of that excited-state character arises from degenerate (paramagnetic) states. As a result of the mixing, the ground $^1A_{1g}$ state remains non-degenerate (and so cannot split) but decreases in energy by an amount proportional to the magnetic perturbation squared and inversely proportional to the energy separation between the admixed states. All the molecules populate this unique ground state and the system has acquired a lower energy by exposure to the applied field. It is therefore paramagnetic. However, because the ground state is unique there can be no change of thermal distribution amongst levels as the temperature is changed and so the paramagnetic susceptibility is independent of temperature.

## 5.5 Orbital Contributions at the Strong-Field Limit

Our discussions of orbital contributions to magnetic moments began with the simple rules for $A$, $E$ and $T$ terms in the weak-field limit. Analogous rules can be constructed when we consider ions in terms of their strong-field configurations. We already had an example with the $t_{2g}^6$ configuration of octahedral cobalt(III) above. An orbital contribution will be made when there is an orbital degeneracy. In the configurations $t_{2g}^1$, $t_{2g}^2$, $t_{2g}^4$, $t_{2g}^5$ there exists the respective three-fold spatial degeneracies given below.

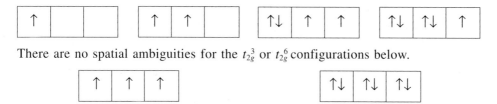

There are no spatial ambiguities for the $t_{2g}^3$ or $t_{2g}^6$ configurations below.

For the cubic-field (octahedral or tetrahedral) subshells $e_{(g)}$, there is spatial degeneracy for $e_{(g)}^1$ but not for $e_{(g)}^2$. Nevertheless, neither of these configurations give rise to an orbital contribution to the magnetic moment. The conditions for orbital contributions to arise in strong-field configurations are that the orbitals must be

**Box 5-8**

*Orbital angular momentum associated with the $d_{xz}$ and $d_{yz}$ orbitals **as a pair**.*
The orbitals $d_{xz}$ and $d_{yz}$ can be expressed in terms of the complex forms $d_1$ and $d_{-1}$ whose angular parts are given by the spherical harmonics $Y_1^2$ and $Y_{-1}^2$, respectively. The matrix of orbital angular momentum about the $z$ axis in the complex basis is

$$
\begin{array}{c|cc}
\hat{l}_z & d_1 & d_{-1} \\
\hline
d_1 & 1 & 0 \\
d_{-1} & 0 & -1
\end{array}
\tag{i}
$$

and we observe the obvious result that $d_1$ and $d_{-1}$ orbitals have $\pm 1$ unit of orbital angular momentum about the $z$ axis. In the real orbital basis, where

$$
d_{xz} = \frac{i}{\sqrt{2}}(d_{-1} + d_1)
$$

$$
d_{yz} = \frac{1}{\sqrt{2}}(d_{-1} - d_1)
\tag{ii}
$$

the equivalent matrix under $\hat{l}_z$ is

$$
\begin{array}{c|cc}
\hat{l}_z & d_{xz} & d_{yz} \\
\hline
d_{xz} & 0 & -i \\
d_{yz} & i & 0
\end{array}
\tag{iii}
$$

On diagonalization, we find the real $d$ orbitals to possess $\pm 1$ unit of orbital angular momentum about $z$, *when taken as a pair*. Although the matrices (i) and (iii) tell the same story, one can barely draw the complex orbitals $d_{\pm 1}$ yet their angular momentum is obvious. If we prefer to use the real forms in (ii), which we can draw, the orbital angular momentum is hidden in the imaginary off-diagonal elements of (iii). These off-diagonal elements have the form

$$
<d_{xz}|\hat{l}_z|d_{yz}> = <d_{xz}|-i|d_{xz}>
\tag{iv}
$$

or, in other words, $l_z$ rotates $d_{yz}$ into $d_{xz}$ (and multiplies it by $-i$). Now the combinations (ii) (or, conversely, $d_{\pm 1}$ from $d_{xz}$, $d_{yz}$) can only be constructed if they are degenerate. All these features, fully discussed for $p_x$ and $p_y$ orbitals in 'Orbitals, Terms and States', are encapsulated within the rules given in the main text above.

degenerate, that the degenerate subsets must be neither full, empty nor exactly half-full, and that at least two of the orbitals within a subset must be related by rotation about the $z$ axis. Within the $t_2$ subset, $d_{xz}$ becomes $d_{yz}$ on rotation about $z$ and so gives rise to an orbital contribution. On the other hand, $d_{z^2}$ and $d_{x^2-y^2}$ of the $e$ subset are not related by rotation about $z$ and give no such orbital contribution. They form the so-called 'non-magnetic doublet' discussed earlier. Although these conditions for orbital magnetic contributions are consistent with the summary given earlier in this section, they must surely seem no more than a magical recipe. For those familiar with the real and complex forms of wavefunctions, a more satisfactory explanation of these rules is given in Box 5-8. Exemplifying the use of the rules in application to the strong-field octahedral configurations of Fig. 5-1, for example, we note that orbital contributions are expected for $d^1$, $d^2$, low-spin $d^4$, low spin $d^5$, high-spin $d^6$, low-spin $d^7$, and $d^9$ configurations.

## 5.6 The Chemical Relevance of Departures from the Spin-Only Formula

Careful and detailed studies of '$d-d$' spectra and magnetic susceptibilities, preferably on samples in the form of single crystals, can yield considerable insight into the bonding in transition-metal complexes. One thinks here of the various specialist techniques that are part of contemporary research which, of course, form no part of our brief in this book. It is the case, however, that our somewhat technical discussions of orbital contributions, second-order Zeeman effects and the like, *are* of direct relevance to the simple exploitation of paramagnetism in inorganic chemistry. They were recognized to be such even in the early days of crystal-field theory and magnetochemistry.

As we have seen, magnetism may be exploited – strictly through the spin-only formula – to count the number of unpaired electrons in a complex and thence to infer something of the nature of the bonding in that complex. This early idea is still employed today at a 'finger printing' level. Within the context of our opening remarks in Section 5.1, a count of unpaired spin can differentiate between strong-field and weak-field environments. These unpaired-electron counting games can only work so simply, however, if we have a strong correlation between the measured property of susceptibility (or effective magnetic moment) and the number of unpaired electrons. The 'spin-only' formula in Eq. (5.11) offers that simple connection. Like so much else in chemical theory, however, it only works when it works! Our discussions in the past few sections show that the formula is sometimes excellent, often reasonably accurate, but just as often inadequate. As such, the simplicity offered by the spin-only relationship is spoilt. Since we do understand why and when, however, as described briefly above, simplicity may have been lost but not understanding.

These remarks are made, therefore, to assure the reader that connections between the number of unpaired electrons in a complex and its magnetic properties – and indeed much more detail – are perfectly possible and well understood, notwithstanding the necessarily brief review of the subject that has been possible in the present non-specialist text.

## 5.7 Summary

In crystal fields of any symmetry, ambiguities can, but need not, arise in assigning the spin-degeneracy of the ground configuration. Such ambiguities arise from the conflict between the desire of electrons to avoid each other versus their desire to avoid negatively charged regions in the environment. In octahedral species, the spin-degeneracy is determined by the relative size of $\Delta_{oct}$ and the mean pairing energy $P$. The crystal-field splitting in tetrahedral complexes of the first row transition-metal complexes is never greater than the pairing energy so that, empirically, only high-spin tetrahedral complexes are observed. The spin-degeneracy of actual systems may be determined from measurements of the magnetic moment. The simplest relationship between spin and magnetism is the limiting case of the spin-only formula relating magnetic moment directly to the number of unpaired electrons. More careful scrutiny of crystal fields and magnetism provides a 'second tier' of sophistication in which departures from the 'spin-only' formula can be anticipated. Orbital contributions to magnetism – which is the antithesis of orbital quenching – are of second order for ions with $A$ or $E$ ground terms, but of first order and complicated for $T$ ground terms. Once again, these qualitative rules derive essentially from symmetry. The details of the physical nature and origins of crystal-field splittings are irrelevant for their establishment. Similar remarks have been made at the conclusions of each of the last three chapters. It is time now to come to grips with the more quantitative side of crystal-field theory and to correlate its successes with other notions of chemical bonding.

## Suggestions for further reading

1. B. N. Figgis, *Introduction to Ligand Fields* , Wiley, New York, **1966**, Chapter 10.
2. F. E. Mabbs, D. J. Machin, *Magnetism and Transition Metal Complexes*, Chapman and Hall, London, **1973**.
3. M. Gerloch, *Magnetism and Ligand-Field Analysis*, Cambridge University Press, Cambridge, **1983**.
   – This is a technical research book on these topics.

# 6 Ligand Fields, Bonding and the Valence Shell

## 6.1 The Nephelauxetic Effect

We saw in Chapter 3 how three spin-allowed transitions arise in octahedral or tetrahedral complexes of metals with $d^2$, $d^3$, $d^7$ or $d^8$ configurations. We also learned that the energies of those transitions depend upon the magnitudes of the crystal field splitting parameter, $10Dq$ and of the interelectron repulsion between the $d$ electrons themselves. One might suppose that while $10Dq$ measures the strength of the interaction between the metal $d$ electrons and their ligand environment, the interelectron repulsion is merely a property of the metal itself. That is not so, however, for the parameter $B$ measures the $d-d$ interactions in the metal *within its particular environment*. Interelectron repulsion energies are every bit as much a probe of the molecular environment of a metal ion as are crystal-field energies. For any given metal complex with a $d^2$, $d^3$, $d^7$ or $d^8$ configuration[*], careful analysis of experimental transition energies yields values of both $B$ and $Dq$. Such analyses have been performed for scores, if not hundreds, of transition-metal spectra. So far as the $B$ parameters are concerned, two general observations have emerged: a) $B$ values for metal complexes are smaller than the values $B_0$ for the corresponding free ions and b) $B$ values may be placed in essentially fixed orders related to ligand or metal.

First, observation a), often written as

$$\beta = B/B_0 < 1 \tag{6.1}$$

and called the nephelauxetic effect, expresses the fact that the repulsions between the $d$ electrons in a complex are less than those in the corresponding free ion. This implies that the average distance between the $d$ electrons in a complex is larger than the average for the corresponding free ion.

Filling out observation b), it is found that for a series of complexes with a common metal, the nephelauxetic effect increases in the order given in Eq. (6.2)

$$\text{nephelauxetic effect increasing} \rightarrow$$
$$\text{F}^- < \text{H}_2\text{O} < \text{NH}_3 < \text{en} < \text{ox}^{2-} < N\text{CS}^- < \text{Cl}^- < \text{CN}^- < \text{Br}^- < \text{I}^- \tag{6.2}$$
$$B/B_0 \text{ decreasing} \rightarrow$$

---

[*] Again, we restrict discussion to spin-allowed transitions here. In general, of course, crystal field effects compete with interelectron repulsion for all $d^n$ configurations, except for $n = 1$ or 9.

The ordering of ligands in Eq. (6.2) is about the same, regardless of the central metal. Analogously for a series of complexes with a common set of ligands, the nephelauxetic effect increases in the order

$$\text{nephelauxetic effect increasing} \rightarrow$$
$$\text{Mn(II)} < \text{Ni(II)} \approx \text{Co(II)} < \text{Mo(II)} < \text{Re(IV)} < \text{Fe(III)} < \text{Ir(III)} < \text{Co(III)} < \text{Mn(IV)}$$
$$B/B_0 \text{ decreasing} \rightarrow \tag{6.3}$$

and, again, the ordering of metals is roughly independent of the ligand set. In fact, it is possible to present the nephelauxetic effect, very roughly, as a simple multiplicative function of independent metal and ligand parameters (Eq. 6.4).

$$(B_0 - B)/B_0 = (1 - \beta) \approx h(\text{ligands}) \times k(\text{metal}) \tag{6.4}$$

**Table 6-1.** Some typical $h$ and $k$ values.

| Ion | $k$ | Ligand | $h$ |
|-----|-----|--------|-----|
| Co(II) | 0.24 | 6 Br⁻ | 2.3 |
| Co(III) | 0.35 | 6 Cl⁻ | 2.0 |
| Cr(III) | 0.21 | 6 CN⁻ | 2.0 |
| Fe(III) | 0.24 | 3 en | 1.5 |
| Mn(II) | 0.07 | 6 F⁻ | 0.8 |
| Ni(II) | 0.12 | 6 H₂O | 1.0 |
| V(II) | 0.08 | 6 NH₃ | 1.4 |

---

**Box 6-1**

*Examples*: a) Using the values in Table 6-1, we find the nephelauxetic reduction for $[NiF_6]^{4-}$ to be $0.8 \times 0.12 = 0.096$, that is, $B/B_0 = 0.904$ or that $B$ in this complex is reduced by about 10% relative to $B_0$ for the $Ni^{2+}$ ion.
b) For $[Co(NH_3)_6]^{3+}$, $(1 - \beta) = 1.4 \times 0.35 = 0.49$, that is, $B$ in this complex is about half of that for the $Co^{3+}$ ion.

---

Qualitatively, at least, there is a unifying theme and chemical correlation to be found in these series. Namely, $\beta$ values decrease with increasing reducing power of the ligands and/or increasing oxidizing power of the metal ions. These two statements can be joined to yield the simple result:

$B$ values decrease as negative charge is transferred from the ligands to the metal.

We can understand this powerful generalization directly from our view of the valence shell in Werner-type complexes as laid out in Chapter 2. Recall that *as an extreme limit* for Werner-type species, we consider the metal contribution to the valence shell for the first-row elements as 4*s* and 4*p,* with 3*d* orbitals excluded. So, the bonds holding the complex together are very largely built from the $4s/4p$ metal orbitals together with appropriate ligand orbitals. As we pass from a free ion to a complex, or as we traverse either of the *nephelauxetic series* (6.2) or (6.3), an increasing amount of electron density is donated by the ligands into the $4s$ and/or $4p$ metal orbitals. These metal orbitals are of the more penetrating type as they have subsidiary maxima fairly close to the metal nucleus (Fig. 2-2). Therefore, a small but significant part of the donated ligand electron density enters those penetrating regions. That density lies between the metal nucleus and the bulk of the 'innocent' $3d$ electrons. Consequently, the $3d$ electrons are more shielded from the nuclear charge in complexes than in the corresponding free ions, and more shielded in complexes characterized by greater ligand $\rightarrow$ metal electron donation than those characterized by less. The greater shielding of the $3d$ orbitals results in their being less well bound, and their radial distribution thus grows. The $3d$ orbitals grow more bulky and more diffuse, and the average distance between $d$ electrons increases. Therefore, the average interelectron repulsion energy decreases, the $B/B_0$ values decrease, and the *nephelauxetic effect* increases.

The name nephelauxetic means 'cloud-expanding'. The explanation for it, which we have just reviewed, will be found elsewhere in the literature under the name 'central-field covalency'. The magnitude of the nephelauxetic effect depends upon the metal, its oxidation state, and the ligands bound to it as summarized in Eqs. (6.2) and (6.3) and Table 6-1. For hexaaquo complexes of various first-row metal(II) ions, for example, $B/B_0$ ranges from 0.8 to 0.9; for many iodo or sulfur donor ligands, one finds $B/B_0$ values in the range 0.5 to 0.7; in certain low-spin cobalt(II) compounds, $B/B_0$ values as low as $0.1-0.3$ have been observed. However, because of certain artefacts in the way the $B$ parameter is defined, none of these nephelauxetic effects should be viewed as implying more than modest expansions of the $3d$ electron clouds. Not the least of such considerations is the fact that the interelectron repulsion energy varies inversely with respect to the electron–electron separation so that we get a reciprocal relationship between $B$ and the degree of cloud expansion, rather than a linear one.

## 6.2 The Spectrochemical Series

The magnitude of $Dq$ in any given complex is clearly a direct measure of the interaction between the 'spectral' metal $d$ electrons and their molecular environment. As for the nephelauxetic effect, values of $Dq$ have been collated for a large number of species and found to fit, very approximately, another multiplicative relationship of metal and ligand functions (Eq. 6.5).

$$Dq \sim f(\text{ligands}) \times g(\text{metal}) \qquad (6.5)$$

Again, ligands may be ordered according to the magnitude of $Dq$ roughly independently of the central metal (Eq. 6.6)

$$I^- < Br^- < SCN^- < Cl^- < F^- < O^{2-} \approx OH^- \approx H_2O < NCS^- < NH_3 < CN^- < PR_3 < CO$$
$$Dq \text{ increasing} \rightarrow \qquad\qquad\qquad (6.6)$$

and metals can be placed in an order roughly independent of the ligand set (Eq. 6.7).

$$Mn(\text{II}) < Ni(\text{II}) < Co(\text{II}) < Fe(\text{III}) < Cr(\text{III}) < Co(\text{III}) < Ru(\text{III}) < Mo(\text{III})$$
$$< Rh(\text{III}) < Pd(\text{II}) < Ir(\text{III}) < Pt(\text{IV})$$
$$Dq \text{ increasing} \rightarrow \qquad\qquad\qquad (6.7)$$

These orderings are called the *spectrochemical series*. At a purely empirical level, the collection of $f$ and $g$ values (Eq. 6.5) in Table 6-2 is reasonably adequate to predict $\Delta_{oct}$ values for various metal-ligand combinations that did not define it. These values are also useful in connection with an empirical relationship known as the *the law of average environment*. This law asserts that the splitting parameter $Dq$ for a metal complex with a mixed set of ligands is given by the appropriately weighted average of the corresponding unmixed complexes. For example, from Table 6-2, $\Delta_{oct}$ for $[NiF_6]^{4-}$ is 8010 cm$^{-1}$ and for $[Ni(H_2O)_6]^{2+}$ it is 8900 cm$^{-1}$; the law of average environment predicts that $\Delta_{oct}$ for $[NiF_4(H_2O)_2]^{2-}$ is 8806 cm$^{-1}$.

---

**Box 6-2**

This procedure is *strictly* invalid, of course, since the symmetry of a six-coordinate complex with dissimilar ligands cannot be exactly octahedral. In this case, further splitting of the $d$ orbitals takes place which is not representable by a single splitting parameter like $\Delta_{oct}$. However, if the departure from $O_h$ symmetry is slight, so that spectral bands are broadened rather than split, the law of average environments retains utility.

---

**Table 6-2.** Some typical $f$ and $g$ values.

| Ion | $g$ | Ligand | $f$ |
|---|---|---|---|
| Co(II) | 9.3 | 6 Br$^-$ | 0.76 |
| Co(III) | 19.0 | 6 Cl$^-$ | 0.8 |
| Cr(III) | 17.0 | 6 CN$^-$ | 1.7 |
| Fe(III) | 14.0 | 3 en | 1.28 |
| Mn(II) | 8.5 | 6 F$^-$ | 0.9 |
| Ni(II) | 8.9 | 6 H$_2$O | 1.0 |
| V(II) | 12.3 | 6 NH$_3$ | 1.25 |

An explanation for the nephelauxetic series came readily to hand. Lets see how successfully we can provide one for the spectrochemical series. First, note that $\Delta_{oct}$ values increase with decreasing size of the donor halides:

$$I^- < Br^- < Cl^- < F^-$$
$$\text{decreasing size} \rightarrow \qquad\qquad (6.8)$$
$$\text{increasing } \Delta_{oct} \rightarrow$$

and this seems reasonable in terms of the simple crystal-field model set out in Section 3.1 in which shorter bonds would indeed imply larger values of $\Delta_{oct}$. However, the ability of the crystal-field model to rationalize the spectrochemical series stops right there. Thus, we might also expect orbital splitting energies to vary according to the charge on the ligands. We observe however, from Eq. (6.6) that the negatively charged halogens produce less orbital splitting than neutral water or ammonia ligands. Secondly, note that $H_2O$, $OH^-$ and $O^{2-}$ ligands define very similar $\Delta_{oct}$ values. We then observe from Eq. (6.7) that $\Delta_{oct}$ increases with metal oxidation state (Eq. 6.9).

$$M(II) << M(III) < M(IV)$$
$$\Delta_{oct} \text{ increases} \rightarrow \qquad (6.9)$$

This would not be expected simply on the basis of a crystal-field model, for the $d$ orbitals will contract with increasing positive charge and hence interact less well with the ligand 'point charges'. The modest decreases in bond length as one traverses the series (Eq. 6.9) are unlikely to compensate for, let alone override, the effects of such orbital contraction. Finally, to add to the confusion, we also note from Eq. (6.7) that $\Delta_{oct}$ values increase as we go down the periodic table (Eq. 6.10).

$$3d << 4d < 5d$$
$$\Delta_{oct} \text{ increases} \rightarrow \qquad (6.10)$$

This might be compatible with the electrostatic model in that the radial extensions of $4d$ and $5d$ orbitals are greater than that of $3d$; but then the diffuseness of these orbitals increases along the series in Eq. (6.10) and that would tend to decrease the $\Delta_{oct}$ values.

These and many similar observations made over the years all make it clear that the simple electrostatic basis of the pure crystal-field model utterly fails to provide even a qualitative understanding of the spectrochemical series. This failure in no way casts doubt upon our successful interpretation of the nephelauxetic effect above. This is because *crystal*-field theory is incompatible with our views about bonding in the valence shell. Should the reader object to having been 'led up the garden path' for the past three chapters, he should remember that, notwithstanding the total failure of crystal-field theory to explain the *magnitudes* of the splitting parameters, their patterns and all properties flowing from them are accounted for with extraordinary success by this model. As we shall see, in replacing crystal-field theory it would be stupid in the extreme to 'throw out the baby with the bath water'.

## 6.3  Bonding in Octahedral Complexes

One might well wonder what crystal-field theory has to do with chemical bonding. After all, all interactions between metal and ligands are deemed repulsive: there is no mention of attractive binding forces. In this respect, crystal-field theory is hardly a *chemical* theory at all. To be fair, it was not developed to be one either. The *qualitative* ideas of the approach, being essentially only dependent upon the $d^n$ configuration and molecular symmetry, are, however, quite compatible with bonding theory, as we shortly describe. We shall also see how the *quantitative* aspects of crystal-field theory, as exemplified by the spectrochemical series, are illuminated by a study of the bonding in transition-metal complexes. The *approach* we make is through a consideration of molecular orbitals in octahedral species.

### 6.3.1  Molecular Orbitals in Diatomic Molecules

Let us first briefly review the construction of molecular orbitals in simple diatomic molecules, AB, using the linear combination of atomic orbitals (LCAO) scheme. The end product for the first long row of the periodic table is the well-known diagram in Fig. 6-1. We focus on two broad principles that are exploited in the construction of this diagram: one has to do with symmetry and overlap, the other concerns energies.

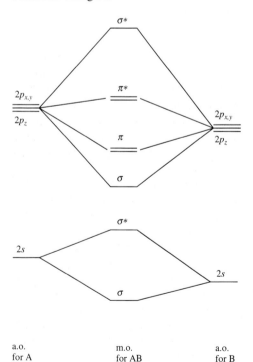

a.o.
for A

m.o.
for AB

a.o.
for B

**Figure 6-1.** Schematic molecular orbital diagram for heteronuclear first-row diatomics.

As to the first, we note the interaction of the $s$ orbital of atom A with the $s$ orbital of B, the $p_z$ with the $p_z$, and the $p_{x,y}$ pair of A with the $p_{x,y}$ pair of B. In principle, of course, we could have considered the possibility of an interaction between, say, the $s$ orbital on A with a $p_x$ orbital on B as shown in Fig. 6-2. The sketch shows that net overlap between these orbitals is zero and so no bonding or antibonding molecular orbitals are formed in this way. Now the labels $s$ and $p_x$ here

in-phase positive
overlap

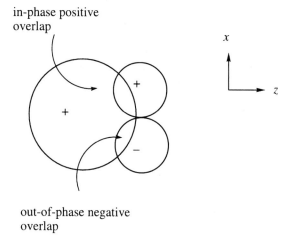

out-of-phase negative
overlap

**Figure 6-2.** Zero net overlap between $s$ ($\sigma$) and $p_x$ ($\pi_x$) orbitals.

are symmetry labels for free atoms. Had we characterized these same orbitals with respect to the cylindrical symmetry of the molecule to be formed, the $s$ orbital would be labelled $\sigma$, and the $p_x$ together with its partner $p_y$, would be labelled $\pi$. Then, using the rule that orbitals of different symmetry do not overlap, our conclusion about the nonbonding interaction between these orbitals follows immediately. While all this is well-known, and almost trivial in the present example, the *classification of fragment orbitals according to the symmetry of the molecule to be formed* gains considerable utility in more complicated systems.

Turning to the second point about energies, we recall how the stabilization of the bonding molecular orbital with respect to the lower of the two atomic orbitals, and the destabilization of the antibonding one with respect to the higher lying atomic orbital, depend upon the magnitude of the relevant overlaps between the interacting atomic orbitals and upon their starting energies. Large bonding and antibonding energy shifts are favoured by large overlap and/or good energy matching between the relevant atomic orbitals. The smaller $\pi-\pi^*$ energy gap relative to the $\sigma-\sigma^*$ (from $p_z - p_z$ overlap) in our example is expected in view of the less good 'sideways' overlap of two $p_x$ orbitals relative to the 'head-on' overlap of two $p_z$ orbitals.

We shall shortly draw on both of these symmetry and energy aspects of Fig. 6-1 in the construction of molecular orbitals for the octahedron. First, however, let us extend the picture to molecules with more than two atoms.

### 6.3.2  Molecular Orbitals in Polyatomic Molecules

Figure 6-1 is a typical molecular orbital diagram in that molecular orbitals in the middle are shown as arising from atomic orbitals on the left and right. The question arises as to what an equivalent diagram might look like for, say, a triatomic molecule. Should we construct one in three dimensions with atomic orbitals of A shown on the left, those of B on the right, and those of C behind? It is possible. But what do we do for a molecule with seven atoms as in an $ML_6$ octahedral complex, for example? We could explore seven-dimensional diagrams! Well we don't do that. Instead, we consider the molecule as notionally broken into fragments which we then consider two at a time. For a triatomic system ABC – linear or not – we could proceed by one of three routes: 1) combine atomic orbitals of A and B to form fragment orbitals for the moiety A – B, and then combine these with atomic orbitals of C to arrive finally at molecular orbitals for the complete ABC molecule, 2) combine B with C to get BC, followed by final combination with the atomic orbitals of A or 3) combine A with C to form AC, and thence include B to get the same final result. A natural question at this stage is to ask whether one particular route is better than the others. But then, what is meant by 'better'? Presumably the best route is that which makes the most of any molecular symmetry and so yields the required result in the shortest time. We illustrate this idea in the following construction of molecular orbitals for the water molecule.

### 6.3.3  Molecular Orbitals for the Water Molecule*

The water molecule possesses two mirror planes of symmetry, as shown in Fig. 6-3. One mirror plane lies in the plane of the diagram through which the whole molecule reflects into itself across the plane. The other, through the oxygen nucleus in the $yz$ plane of the figure, and shown by the dotted line, reflects $H_a$ into $H_b$ and vice versa.

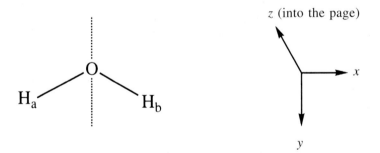

**Figure 6-3.** One mirror plane is the plane of the page, the other (----) is normal to the page.

---

* A fuller and interesting account of molecular ( and of 'equivalent') orbitals will be found in Murrell, Kettle, Tedder, *Valency Theory*, 2nd ed., Wiley, New York, **1969**, p 190.

We construct molecular orbitals for the complete $H_2O$ molecule, by first considering combinations of the symmetrically related hydrogen atomic orbitals, and then by combining these with the oxygen atomic orbitals. We proceed by three steps: (A) combine the H $1s$ orbitals and classify them according to their behaviour with respect to the mirror planes discussed above, (B) classify the oxygen atomic orbitals with respect to the same molecular symmetry, and (C) form $H_2O$ bonding and antibonding molecular orbitals by overlap of H···H and O orbitals of matching symmetry.

*Step (A):* Form combinations of H $1s$ orbitals

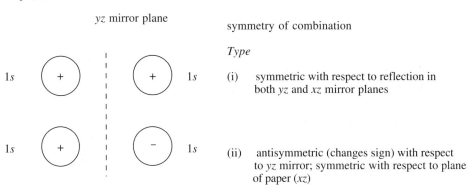

*Step (B)*: Classify O atomic orbitals

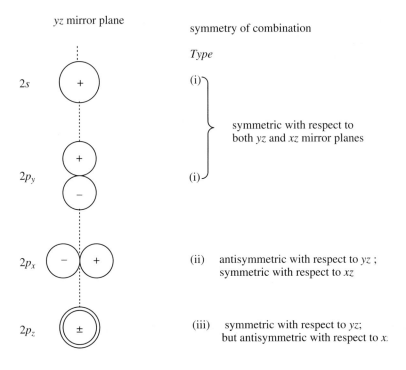

*Step (C)*: Combine like with like

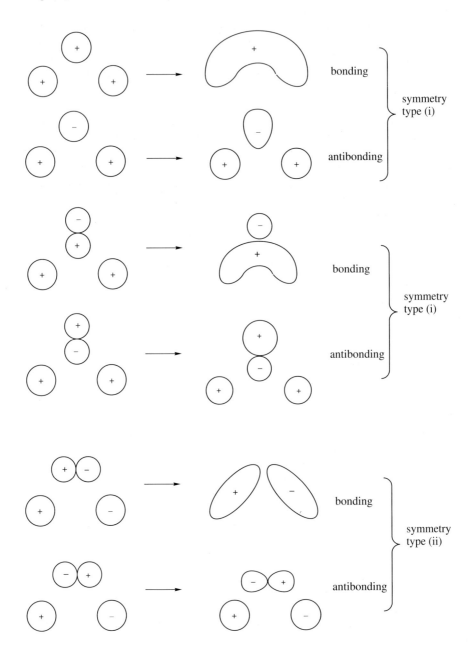

The hydrogen orbitals do not form a combination of symmetry type (iii) and so leave the oxygen $p_z$ orbital as nonbonding.

To complete the full molecular orbital diagram we should consider energies (determined in part by relevant overlap integrals) and the possibility of mixing between the molecular orbitals shown above as of like symmetry. However, we need not bother with all that here. The purpose of this exercise has been to introduce the concept of *group orbitals*. In our example, we constructed two such group orbitals – those in step (A) showing the constructive and destructive combinations of H $1s$ atomic orbitals. It is of no consequence that the magnitude of the overlap between these orbitals on well separated atoms is very small. We may still consider a combination like $[\psi_1(H_{1s}) + \psi_2(H_{1s})]/\sqrt{2}$ to be a group orbital. Then, as we have seen in step (C), these group orbitals are combined with symmetry matching oxygen atomic orbitals. Ignoring questions of relative energies and other quantitative matters, a final molecular orbital diagram (Fig. 6-4) may be constructed in a similar fashion to that shown for the diatomic case.

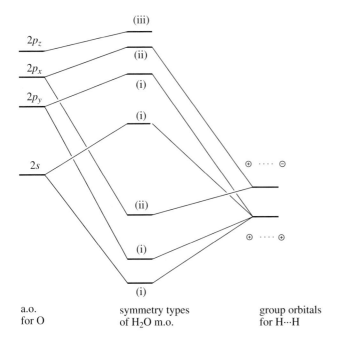

a.o.                 symmetry types              group orbitals
for O                of H$_2$O m.o.              for H···H

**Figure 6-4.** Schematic molecular orbital diagram for water.

We are now ready to apply the ideas in the preceding three sections to the construction of molecular orbitals in octahedral complexes.

### 6.3.4  The Molecular Orbital Diagram for Octahedral Complexes: Local M–L $\sigma$ Bonding

We begin by restricting consideration of the bonding in octahedral complexes to interactions between the metal and any one ligand but involving only local $\sigma$ orbitals. By this we mean that we imagine each ligand to have only $s$ or metal-directed $p$ orbitals (or both) available for overlap with the metal. We label the available orbital on ligand $i$ as $\sigma_i$. Further, and important, complications that arise when we include local ligand $\pi$ orbitals into our scheme, are discussed in Section 6.3.6.

With this restriction to only local M–L $\sigma$ bonding, we proceed as we did for the water molecule above. Just as the two hydrogen atoms are there spatially related by the molecular mirror symmetry, so here are the six ligands related by the molecular octahedral symmetry. So, in step (A), we combine the ligand orbitals and classify the resulting group orbitals according to the octahedral symmetry. Then, in step (B), we label the metal atomic orbitals according to that same symmetry, and finally, in step (C), we combine like with like.

The results of pursuing steps (A), (B) and (C) are given in Table 6-3 and Figs. 6-5, 6-6 and 6-7. Let us consider them in detail. In Fig. 6-5 we see how the combination of ligand orbitals, $(\sigma_1 + \sigma_2 + \sigma_3 + \sigma_4 + \sigma_5 + \sigma_6)/\sqrt{6}$, is symmetry matched to the central metal $s$ orbital (the factor $\sqrt{6}$ is included simply to normalize this group

**Table 6-3.** Labelling of metal and ligand group orbitals in $O_h$ symmetry.

| Symmetry | Metal orbital | Ligand group orbital |
|---|---|---|
| $a_{1g}$ | $s$ | $1/\sqrt{6}\,(\sigma_1 + \sigma_2 + \sigma_3 + \sigma_4 + \sigma_5 + \sigma_6)$ |
| $e_g$ | $d_{z^2}$ | $1/\sqrt{12}\,(2\sigma_3 + 2\sigma_6 - \sigma_1 - \sigma_2 - \sigma_4 - \sigma_5)$ |
|  | $d_{x^2-y^2}$ | $1/2\,(\sigma_1 - \sigma_2 + \sigma_4 - \sigma_5)$ |
| $t_{1u}$ | $p_x$ | $1/\sqrt{2}\,(\sigma_1 - \sigma_4)$ |
|  | $p_y$ | $1/\sqrt{2}\,(\sigma_2 - \sigma_5)$ |
|  | $p_z$ | $1/\sqrt{2}\,(\sigma_3 - \sigma_6)$ |

orbital). The appropriate symmetry label for this combination in octahedral $(O_h)$ parity is $a_{1g}$. Recall our use of these labels in Chapter 3: $a$ means one-fold spatial degeneracy, $g$ means even (symmetric) under inversion through the centre of symmetry. In combining the metal $s$ orbital with this ligand group orbital, we construct bonding (b) or antibonding (a) molecular orbitals for the complete $ML_6$ complex, depending upon whether the two orbitals (metal and ligand group) are in phase or out of phase:

$$\psi^b{}_{mo}(a_{1g}) = a_1\chi(s) + b_1(\sigma_1 + \sigma_2 + \sigma_3 + \sigma_4 + \sigma_5 + \sigma_6)$$
$$\psi^a{}_{mo}(a_{1g}) = a'_1\chi(s) - b'_1(\sigma_1 + \sigma_2 + \sigma_3 + \sigma_4 + \sigma_5 + \sigma_6) \qquad (6.11)$$

where $a_1$, $a'_1$, $b_1$, $b'_1$ are all positive. (The reason why $a_1 \neq a'_1$ and $b_1 \neq b'_1$ derives from the inclusion of non-zero overlap integrals between metal and ligand group orbitals).

The three-fold degenerate set of $p$ orbitals are labelled $t_{1u}$ ($t$ for three-fold, u for odd under inversion through the centre of symmetry). As shown in Fig. 6-6, each metal $p$ orbital matches symmetry with ligand group orbitals comprising just two

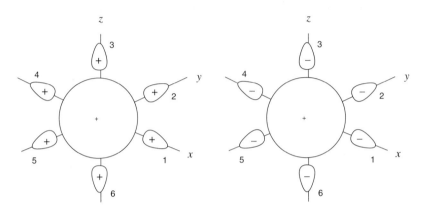

bonding $a_{1g}$ interaction          antibonding $a_{1g}$ interaction

**Figure 6-5.** Interaction with the metal **s** orbital.

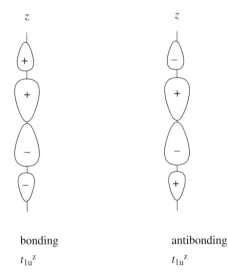

bonding          antibonding

$t_{1u}{}^z$          $t_{1u}{}^z$

**Figure 6-6.** Bonding and antibonding $t_{1u}^x$, $t_{1u}^y$ combinations are similar but with orbital densities along the $x$ or $y$ axes respectively.

out of the six available ligand $\sigma$ functions. Bonding and antibonding $t_{1u}$ molecular orbitals for the octahedral molecule are then formed as

$$\psi^b_{mo}(t_{1u}^x) = a_2\chi(p_x) + b_2(\sigma_1 - \sigma_4)$$
$$\psi^a_{mo}(t_{1u}^x) = a_2'\chi(p_x) - b_2'(\sigma_1 - \sigma_4)$$

(6.12)

together with equivalent (and degenerate) combinations between $p_y$ with $(\sigma_2 - \sigma_5)$ and between $p_z$ with $(\sigma_3 - \sigma_6)$.

In Fig. 6-7, similar procedures are followed for the metal $d$ orbitals. The $d_{x^2-y^2}$ orbital-symmetry matches with the in-plane group combination $(\sigma_1 - \sigma_2 + \sigma_4 - \sigma_5)/2$ to give the molecular orbitals described in Eq. (6-13).

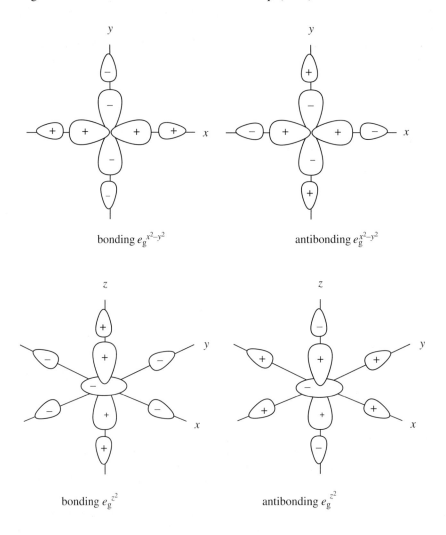

bonding $e_g^{x^2-y^2}$          antibonding $e_g^{x^2-y^2}$

bonding $e_g^{z^2}$          antibonding $e_g^{z^2}$

**Figure 6-7.** Interaction with the metal $d_{x^2-y^2}$ and $d_{z^2}$ orbitals.

$$\psi_{mo}^{b}(e_g^{x^2-y^2}) = a_3\chi(d_{x^2-y^2}) + b_3(\sigma_1 - \sigma_2 + \sigma_4 - \sigma_5)$$
$$\psi_{mo}^{a}(e_g^{x^2-y^2}) = a_3'\chi(d_{x^2-y^2}) - b_3'(\sigma_1 - \sigma_2 + \sigma_4 - \sigma_5)$$

(6.13)

The different shape of the $d_{z^2}$ orbital is matched by the ligand group orbital $(2\sigma_3 + 2\sigma_6 - \sigma_1 - \sigma_2 - \sigma_4 - \sigma_5)/\sqrt{12}$, and we get the molecular orbitals in Eq. (6.14).

$$\psi_{mo}^{b}(e_g^{z^2}) = a_4\chi d_{z^2}) + b_4(2\sigma_3 + 2\sigma_6 - \sigma_1 - \sigma_2 - \sigma_4 - \sigma_5)$$
$$\psi_{mo}^{a}(e_g^{z^2}) = a_4'\chi(d_{z^2}) - b_4'(2\sigma_3 + 2\sigma_6 - \sigma_1 - \sigma_2 - \sigma_4 - \sigma_5)$$

(6.14)

Finally, note that no combination of ligand $\sigma$ orbitals interacts with members of the metal $t_{2g}$ set. The vanishing overlap between any ligand $\sigma$ orbital and, say, the $d_{xy}$ orbital is illustrated in Fig. 6-8. Overall, therefore, the metal $t_{2g}$ orbitals are non-bonding in this scheme. Recall how the $2p_z$ orbital of oxygen is similarly nonbonding to the hydrogen orbitals in water.

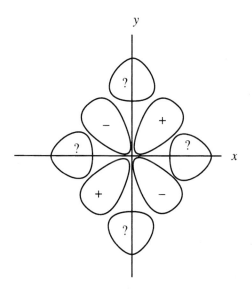

**Figure 6-8.** Impossible symmetry matching of ligand $\sigma$ orbitals with metal $d_{xy}$. Similar results apply for the $xz$ and $yz$ planes.

We now collect together the various parts illustrated in Figs. 6-5 – 6.7 by making some simple assumptions about the relative magnitudes of these metal-ligand interactions. In this we refer to the arguments given in Section 2.2, namely, that the bonding interaction between the metal $4s$ orbital and the ligands will be greater than that for the metal $4p$ and that, because of radial compactness, the metal $3d$ orbitals will form the weakest interactions of all. Qualitatively, therefore, the complete molecular orbital diagram for a first-row octahedral complex with only local $M-L$ $\sigma$ interactions is expected to like that in Fig. 6-9.

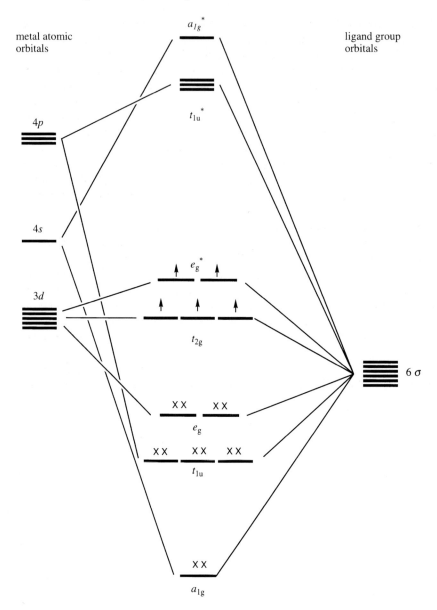

**Figure 6-9.** Schematic m.o. diagram for metal and $\sigma$-bonding ligands in $O_h$ symmetry.

The final step is to house the electrons. Each ligand, acting as a $\sigma$ donor, is considered to offer a lone pair of electrons. This is the case regardless of whether the ligand is formally negatively charged, like $Cl^-$, or neutral like $NH_3$. Ultimately, these electrons are *shared* by the metal and ligand so that there is no implication, now or later, that the ligand 'donates away' or 'loses' two negative charges. In addition to these twelve electrons are $n$ more, originating from the $d^n$ configuration of the metal, giving $(12 + n)$ in all. Applying the Aufbau principle, we place twelve of these electrons in the lowest six bonding molecular orbitals, $a_{1g}$, $t_{1u}$ and $e_g$, as indicated in the figure by crosses. The remaining $n$ electrons then occupy the nonbonding $t_{2g}$ orbitals and the antibonding $e_g^*$ orbitals. We differentiate between the first twelve and the remaining $n$ electrons by crosses and arrows, not because these electrons are different in any way – for, of course, they are not – but to emphasize the connections between the present molecular orbital approach on the one hand, and the ideas of crystal-field theory on the other. Thus, the lowest six filled molecular orbitals provide an account of the binding, attractive forces between the metal and the ligands. Above them lie first the $t_{2g}$ and then the (antibonding) $e_g^*$ orbitals, amongst which are distributed the same number of electrons as defined by the metal $d^n$ configuration. In short, we may *map* this latter distribution onto that discussed so fully earlier under the heading 'crystal-field theory'.

Crystal-field theory accounts for the $t_{2g}-e_g$ splitting, $\Delta_{oct}$, in terms of the differential *repulsion* of the various electrons by ligands viewed as point charges. Within the molecular orbital scheme, on the other hand, that splitting is seen in terms of the antibonding energy of the $e_g^*$ molecular orbital (and of the $t_{2g}$, as we shall see in the next section). In turn, larger antibonding (repulsive) energies for the $e_g^*$ molecular orbitals are to be associated with larger bonding (attractive) energies for the $e_g$ molecular orbitals. Antibonding interactions are repulsive, as bonding ones are attractive.

Immediately, therefore, we have one insight into the spectrochemical series by noting that both bonding and antibonding energy shifts of $e_g$ and $e_g^*$ orbitals are

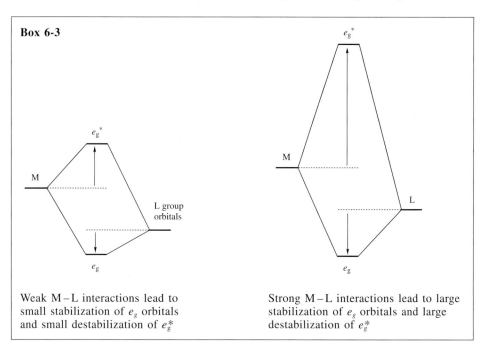

**Box 6-3**

Weak M–L interactions lead to small stabilization of $e_g$ orbitals and small destabilization of $e_g^*$

Strong M–L interactions lead to large stabilization of $e_g$ orbitals and large destabilization of $e_g^*$

closely related to the donor ability of the ligands in the given complex. However, we need to do more work yet before we can construct any really successful account of the spectrochemical series. We will return to this question shortly. Furthermore, we warn the reader that the immediately preceding discussion is somewhat naive and must be re-addressed later (see Section 6.4).

### 6.3.5 Charge-Transfer Transitions

The '$d-d$' bands involve electronic rearrangements within the $t_{2g}$ and $e_g$ (now $e_g^*$) orbitals. The so-called 'charge-transfer' bands, on the other hand, involve transitions between one or the other of these '$d$' orbital subsets and other molecular orbitals in Fig. 6-9. Some of these concern the promotion of an electron from the bonding $t_{1u}$ orbitals into the $t_{2g}$ or $e_g$ orbitals, others may concern electronic transitions from the $t_{2g}/e_g$ orbitals into the $t_{1u}^*$ antibonding orbitals. We focus on these two types of transition, rather than $a_{1g} \rightarrow e_g$ etc., as they alone amongst the many promotions possible in Fig. 6-9 involve a parity change: $u \rightarrow g$ or $g \rightarrow u$. Provided other selection rules are satisfied, therefore, such transitions may be fully allowed. It is also obvious from the qualitative ordering of molecular orbitals in the figure that these electronic transitions will occur at higher energies (larger promotion energies) than those of '$d-d$' type transitions. We thus have an explanation for the generalization, exemplified in Section 2.1, that transition-metal spectra often show intense bands at higher energies than the weak '$d-d$' bands. Recall here the discussion about the perceived colours of the chloro- and bromocuprate ions in Section 2.1.

The name 'charge-transfer' arises from the fact that these transitions take place between $t_{2g}$ or $e_g$ orbitals, which are largely of metal character (little mixed with any ligand orbitals), and members of the bonding or antibonding sets that possess very much greater ligand character. Hence, these transitions involve a much greater displacement of charge, one way *or* the other, between metal and ligand than do '$d-d$' transitions.

Finally, we must remember that just as a '$d-d$' spectrum is not properly described at the strong-field limit – that is, without recognition of interelectron repulsion and the Coulomb operator – neither is a full account of the energies or number of charge-transfer bands provided by the present discussion. Just as a configuration $t_{2g}^n e_g^m$ gives rise to several terms, often with different spin, so also do excited configurations like $t_{2g}^{n-1} e_g^n t_{1u}^1*$. So we must expect the charge-transfer spectrum to be every bit as complicated as the '$d-d$'. While we do not pursue this complex matter further in this book, it is always well to keep in mind the fact that molecular orbital diagrams like that in Fig. 6-9 are but the beginning of any bonding picture.

### 6.3.6 Metal-Ligand π Bonding

Each local metal–ligand interaction in a complex might include a contribution from π bonding. We now remove the restriction of only local σ bonding adopted above, and consider the contribution of $\pi$ bonding ligands to our molecular orbital scheme. Important differences arise between ligands as $\pi$ donors or as $\pi$ acceptors and we study these two situations side-by-side. For concreteness' sake, we may envisage the $\pi$ orbitals on either type of ligand as $p$ functions on the ligating atoms directed *normal* to the local M–L vector. Combinations of ligand $\pi$ orbitals form only one group of importance that is symmetry matched to metal orbitals. These group orbitals, of $t_{2g}$ symmetry in the octahedron, comprise a degenerate set of three and are shown in Fig. 6-10 together with the appropriate members of the $t_{2g}$ set of metal $d$ orbitals with which they overlap.

We now incorporate the bonding and antibonding $t_{2g}$ molecular orbitals of Fig. 6-10 into the energy diagram of Fig. 6-9. So as not to obscure the important issues

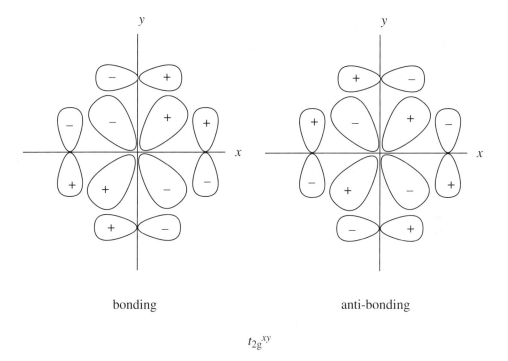

bonding          anti-bonding

$t_{2g}{}^{xy}$

**Figure 6-10.** Symmetry matching of metal $t_{2g}$ orbitals with ligand $\pi$ functions. Similar diagrams may be drawn in the $xz$ and $yz$ planes.

---

**Box 6-4**

Other group orbitals deriving from ligand $\pi$ functions transform as $t_{1u}$, $t_{1g}$, and $t_{2u}$. The $t_{2u}$ group orbital can match the symmetry of some of the metal $f$ orbitals and the $t_{1g}$ matches some metal $g$ orbitals; we are concerned with neither of these here. The $t_{1u}$ ligand group orbital can overlap with the $t_{1u}$ set of metal $p$ orbitals already used for $\sigma$ bonding. Any such overlap is expected to be small, however, involving as it does 'sideways' overlap of metal $p$ and ligand $p$ orbitals on well separated centres. Accordingly, we also neglect this contribution to our bonding scheme, but here only for reasons of simplicity.

that arise, we draw incomplete diagrams, as in Fig. 6-11, in which only the $e_g$ orbitals of the $\sigma$-bonding scheme are made explicit. Two diagrams are presented, one for each of the two cases of ligand $\pi$ donors and $\pi$ acceptors. In Fig. 6-11a, we recognize that the $\pi$ orbitals of $\pi$-donor ligands will be of relatively low energy (like the $\sigma$-donor orbitals) and be filled in the free ligands. Consequently, the bonding octahedral $t_{2g}$ molecular orbitals will also be filled. The crystal-field splitting parameter, $\Delta_{\text{oct}}$, is now identified with the energy gap $t_{2g}^* - e_g^*$. Recall how crystal-field theory considers all $d$ orbitals as repelled by the ligands, but the $e_g$ more than the $t_{2g}$ subset because they point directly at the ligands. In the molecular orbital

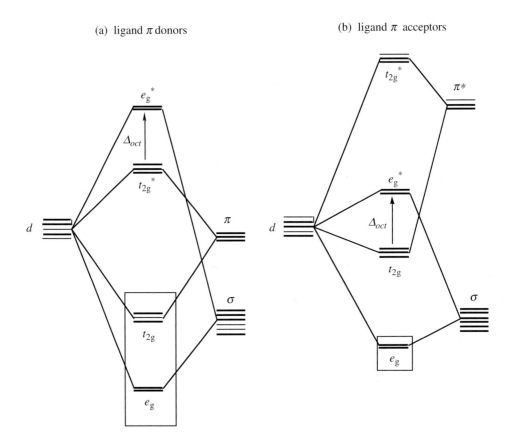

**Figure 6-11.** Metal–ligand $\pi$ bonding in $O_h$ symmetry. The boxed orbitals are filled with electrons – *notionally* from the group of six ligands.

scheme of Fig. 6-11, the antibonding interaction with the $e_g$ orbitals is greater than that with the $t_{2g}$, principally because $\sigma$ overlap is greater than $\pi$. Finally, note that $\Delta_{\text{oct}}$ is less for a ligand which is both a $\sigma$- and $\pi$-donor than for a pure $\sigma$-donor. Greater $\sigma$ bonding increases the energy of the $e_g^*$ orbitals and $\Delta_{\text{oct}}$, while greater $\pi$ bonding increases the energy of the $t_{2g}^*$ orbitals and hence *decreases* $\Delta_{\text{oct}}$. It is these opposing trends that we see as partly responsible for the lower position (smaller $\Delta_{oct}$) of the halogens in the spectrochemical series than that of the $\pi$-neutral ammonia ligand, for example.

In Fig. 6-11b we consider the case of $\pi$-acid (acceptor) ligands. Here, the ligand functions are empty high-energy ligand molecular orbitals we label $\pi^*$. We will look at an example shortly. The stacking of the octahedral molecular orbitals now takes on the different ordering shown in Fig. 6-11b and, because the bonding $t_{2g}$ orbitals are empty (the ligands are $\pi$-*accepting* now), the crystal-field splitting, $\Delta_{oct}$, is associated with the energy gap $t_{2g} - e_g^*$. This time, the greater the $\pi$-acidity of the ligand set, the lower the energy of the bonding $t_{2g}$ orbitals and the greater $\Delta_{oct}$. The position of the CO ligand at the higher end of the spectrochemical series is ascribed to their strong $\pi$-accepting role. Overall, therefore, study of the bonding in octahedral complexes by molecular orbital methods predicts that $\Delta_{oct}$ will be smaller for $\pi$-donor ligands and larger for $\pi$-acceptors, with $\pi$-neutral species inbetween. Broadly, that is what is found experimentally.

## 6.4 Ligand-Field Theory

The discussions of the past few sections are often cited to define ligand-field theory as the application of molecular orbital theory to transition-metal complexes. **It is no such thing.** Certainly we can see how, from a first look at chemical bonding by molecular orbital methods, one should not take a literal, that is pure electrostatic, view of the origins of crystal fields. Apart from those already raised in Section 6.2, there are abundant objections to the simplistic crystal-field view. Numerical computations of the magnitude of $\Delta_{oct}$ within a variety of electrostatic models have yielded widely disparate estimates – some even of the wrong sign. As Van Vleck made clear in 1935, the important conclusion to be drawn from schematic studies like those described above is that the quantity $\Delta_{oct}$ is to be viewed as a *parameter* of the system which, *inter alia*, subsumes the consequences of covalent bonding. *Most importantly*, note too how such bonding could be dominated by non-*d* orbitals just as much as being a roughly equal property of all orbitals.

Let us look a little more closely at this last point. Suppose, for the sake of argument, that we take the extreme viewpoint mentioned in Section 2.2 that the $3d$ orbitals of a first-row transition metal (in higher oxidation states, remember) are so contracted that they effectively do not overlap with the ligand orbitals at all. This would imply that the $t_{2g} - e_g^*$ energy gap in Fig. 6-9 should be vanishingly small. Suppose further that the complex under consideration is one formed with neutral, rather than negatively charged, ligands. Any pure crystal-field (point-charge) splitting of the $t_{2g}$ and $e_g$ orbital subsets would similarly be expected to be very small. Even under these combined circumstances, however, we should still expect to observe a significant $t_{2g} - e_g$ energy splitting. That would arise from the Coulombic interaction of the $d$ electrons with the non-spherical environment. For, don't forget, even if the $3d$ orbitals overlap insignificantly with the ligands, the same is not true of the $4s$

---

**Box 6-5**

By the way, recall the trend in Eq.(6.9) which was at odds with the *crystal*-field premise. Within the *ligand*-field picture, both bonding electron density and (in the limit, nonbonding) *d*-electron density are increasingly drawn in towards the metal nucleus with increasing formal metal charge. These two electron densities are accordingly brought into closer mutual proximity and consequently, $Dq$ values increase along this series.

---

(or $4p$). *That* overlap, and bonding, is what holds the complex together and the bonding electron distribution which results is octahedrally deployed, not spherically. All in all, therefore, we would find the $t_{2g}$ and $e_g$ subsets of the $3d$ electrons energetically differentiated by their differing proximities to the bonding electron density. In effect, the $d$ electrons may be thought of as repelled by the *bonds* rather than by the ligands as point charges. Since the bonding electron density takes large values even in those regions reasonably close to the metal nucleus, the $d$-electron – bonding electron interaction is expected to be significant. Of course, a numerical computation of the $t_{2g}$–$e_g$ splitting would be an extremely complex affair, not least because it would be predicated on a full all-electron calculation of the bonding itself. Nevertheless, even without such a calculation it is apparent that the quantity we call $\Delta_{\mathrm{oct}}$, though reflecting the underlying bonding in a given complex, must not be naively thought of as simply – perhaps linearly – related to the $t_{2g}$–$e_g$ splitting that one might compute in a construction of a *schematic* diagram like Fig. 6-9. Note too that any differences between the schematic molecular orbital diagram and a full all-electron calculation (the latter is, in any case, virtually impossible with current computational facilities) are likely to be large with respect to the scale of the ligand-field energies.

So, *ligand-field theory* is the name given to crystal-field theory that is freely *parameterized*. The centrally important point is that ligand-field calculations, whether numerical or merely qualitative, explicitly or implicitly employ a ligand-field Hamiltonian, very much like the crystal-field Hamiltonian, operating upon a basis set of *pure d* orbitals. Instead of the crystal-field Hamiltonian (Eq. 6.15),

$$\mathcal{H}_{CF} = \sum_{i<j}^{n} \frac{e^2}{r_{ij}} + V_{CF} \tag{6.15}$$

in which $V_{CF}$ takes a form describing the potential energy established by an array of point charges, for example, we use the ligand-field Hamiltonian (Eq. 6.16),

$$\mathcal{H}_{LF} = \sum_{i<j}^{n} U(i,j) + V_{LF} \tag{6.16}$$

in which $V_{LF}$ is an *effective* one-electron operator called the *ligand-field potential*. Note also that interelectron repulsions are no longer calculated with the explicit Coulomb operator, as in Eq. (6.15), but by an effective, two-electron operator $U(i,j)$. This change is commensurate with the nephelauxetic effect, in which free-ion $B_o$ values are replaced with $B$ values in the complex, in recognition of electron density changes brought about by covalency.

---

**Box 6-6**

A simple example of an 'effective' operator with which the reader will be familiar is the use of $Z_{\text{eff}} \, e/r$ as the effective nuclear potential experienced by an electron outside of a closed inner shell. Thus, we may compute the energies and wavefunctions for a $2s$ or $2p$ electron outside a $1s^2$ shell, using the 'hydrogen-like' Hamiltonian,

$$\mathcal{H}_{H-like} = -\frac{1}{2}\nabla^2 - \frac{Z_{eff}e^2}{r}$$

but note that the value of $Z_{\text{eff}}$ is different for an outer $2s$ electron compared with a $2p$ electron. Well-known discussions of this difference centre upon the concept of variable shielding and orbital penetration, of course.

---

It is no part of our thesis in this book to get too technical. At the same time, however, it is surely unacceptable that a qualitative approach should avoid all mention of technicalities if that tactic results in a complete misunderstanding of the quite different standings of ligand-field theory on the one hand, and molecular-orbital theory on the other. Molecular orbital based discussions, like those in the immediately preceding sections, provide insight into some of the trends in, and factors affecting, ligand-field parameters. However, these two models do not map onto one another as computational procedures. Furthermore, the ligand-field approach is closer to the end result so far as $d$ electron properties are concerned, than are various conventional molecular orbital schema.

Molecular orbital calculations may employ any convenient basis and, in many-electron applications, those bases will generally include within them some recognition of all kinds of two-electron interactions. Metal $s$, $p$ and $d$ functions, for example, will be treated in this regard on essentially equal footings. In ligand-field calculations, on the other hand, all manipulations are made with respect to a pure $d$ basis, this being defined *only* with respect to the angular momentum property, being given by $l = 2$, with no required statement about the radial part which might otherwise define a free-ion **3**$d$ function. Furthermore, $d-d$ interaction energies are explicitly dealt with separately from all others (so we have separate parameters, like $B$ for '$d-d$' interelectron repulsions on the one hand, and $\Delta_{oct}$ for the ligand-field on the other). One is to realize that it is, *a priori*, strange and unexpected to find that useful calculations of '$d-d$' spectral transition energies and magnetic properties, amongst others, can be performed in this way. The $d$ orbitals are *explicit* in such calculations while all others (metal $s$, $p$; ligand functions) are left *implicit* within the effective operators of Eq. (6.16). Molecular orbital calculations, on the other hand, include all reasonable orbitals explicitly within a chosen basis set and the operators are direct in the sense that they do not contain within them any other basis functions. Molecular orbital calculations are thus formally able to provide quantitative accounts of all molecular properties *ab initio* while ligand-field theory relates only to $d$-electron properties, and then only in a parametric fashion. From this remark, it might appear that the molecular orbital technique is clearly the superior. However, within the domain of transition-metal '$d-d$' spectra and magnetism, the *fact* is that, while ligand-field theory uniformly and consistently provides a quantitative account of experiment together with much insight into the underlying chemical bonding, corresponding molecular orbital calculations are today too complex to compete at all in practice. It is equally evident that empirical molecular orbital methods are also ill-suited to this special range of properties.

There is a parallel to ligand-field theory elsewhere in chemistry. The Hückel theory of organic $\pi$ electron systems analogously focuses on just one subset of electrons. While various $\pi$ orbitals are recognized explicitly, the roles of the $\sigma$-bonding framework in such species are kept implicit, being 'folded into' the effective operators of the model. The Hückel approach (and here we mean the 'simple' Hückel model rather than various Extended Hückel approaches) is extraordinarily successful as a parametric model of a particular set of electrons. The successes of Hückel theory and ligand-field theory, strictly within their own domains, is owed to the way in which Nature keeps separate the relevant electronic sets. In $\pi$-electron theory, orthogonality is of the essence; in ligand-field theory, the (less perfect) decoupling of $d$ electrons from all others derives from their relatively contracted nature, as discussed in Section 2.2. Were the $d$ electrons less decoupled, i.e. more admixed with the valence shell, ligand-field theory simply would not work. By 'work', we don't just mean qualitatively, as established perhaps merely by symmetry, but quantitatively. It is unfortunate that the space and level of presentation prevent our justifying and exemplifying that ligand-field theory really does work at a quantitative level: we ask the reader to be assured that it is so.

Finally, on this question of the efficacy of ligand-field theory depending on Nature's selection of a relatively isolated subset of electrons, consider what might be the limits to the domain of this approach. As will be discussed briefly in Chapter 10, the $f$ orbitals in lanthanoide complexes are even more 'buried' beneath the valence shell than are the $d$ electrons in the main transition series. It will be of no surprise, therefore, to learn that ligand-field studies on the spectra and magnetism of lanthanoide complexes, though technically rather more complex than for the $d$ block species, are entirely successful. On the other hand, suggestions which have been made in the literature from time-to-time that ligand-field techniques might be applicable to $p$-block compounds are ill-founded. There is no corresponding set of $p$ electrons that is well isolated from other electrons in those systems. Similarly, ligand-field theory is not applicable to the charge-transfer spectra described in Section 6.3.5 for there we stepped outside of the $d$ shell and included members of the valence shell itself. The above are clear-cut examples of the applicability of ligand-field theory; of the domain or 'regime' of the theory. Less clear areas also come to mind in which the approach may gradually begin to fail. These might include, perhaps, the periphery of the $d$ block, perhaps the third transition series. Paradoxically, and unfortunately, it is difficult to test this proposition since in those systems, charge-transfer spectra frequently obscure the '$d-d$' bands whose analysis might provide the answer. To circumvent this, we might consider complexes in low oxidation states. Indeed we have at last reached that long promised topic.

## 6.5 Synergic Back-Bonding

Just as the statement that 'such-and-such a compound is stable' is meaningless unless one adds 'with respect to' something, so also is the definition of a ligand as a 'good donor'. A dramatic illustration of this idea is provided by carbon monoxide. A molecular orbital diagram for the free CO molecule is shown in Fig. 6-12. We note that the highest occupied molecular orbital (HOMO), $3\sigma$, is a $\sigma$-bonding orbital extending beyond the internuclear vector:

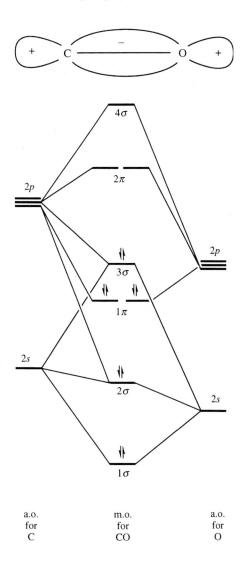

**Figure 6-12.** Schematic molecular orbital diagram for carbon monoxide.

The lobes of electron density outside the C–O vector thus offer σ-donor lone-pair character. Surprisingly, carbon monoxide does not form particularly stable complexes with $BF_3$ or with main group metals such as potassium or magnesium. Yet transition-metal complexes with carbon monoxide are known by the thousand. In all cases, the CO ligands are bound to the metal through the carbon atom and the complexes are called carbonyls. Furthermore, the metals occur most usually in low formal oxidation states. Dewar, Chatt and Duncanson have described a bonding scheme for the metal–CO interaction that successfully accounts for the formation and properties of these transition-metal carbonyls.

We see from Figure 6-12 that free carbon monoxide is bound in both σ and π modes. As oxygen is more electronegative than carbon, resulting in the lower energies of the oxygen atomic orbitals than the carbon ones in the figure, the bonding π molecular orbital, 1π, favours oxygen:

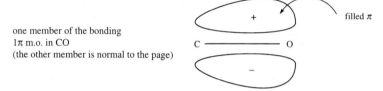

one member of the bonding
1π m.o. in CO
(the other member is normal to the page)

Correspondingly, the (empty) antibonding 2π molecular orbital favours the carbon:

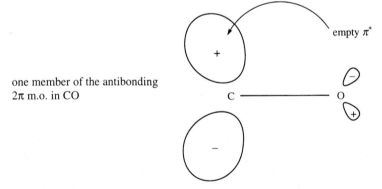

one member of the antibonding
2π m.o. in CO

In low oxidation states, transition metals possess filled or partly filled $d$ shells. The Dewar-Chatt-Duncanson model envisages some of that electron density in (local) $d_\pi$ (e.g. $d_{xz}$, $d_{xy}$) orbitals being donated into the empty π* orbitals of the carbon monoxide:

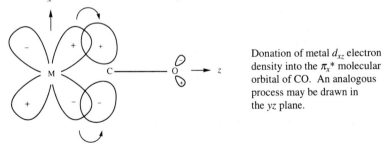

Donation of metal $d_{xz}$ electron density into the $\pi_x$* molecular orbital of CO. An analogous process may be drawn in the $yz$ plane.

Accordingly, the CO moiety acquires negative charge. The consequent exigencies of the electroneutrality principle are then met by the CO group donating this charge back to the metal via its now expanded $\sigma$-donor orbital:

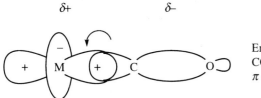

Enhanced $\sigma$ donor ability of the CO towards the enhanced $\pi$ accepting metal.

The effect is cyclic or synergic. The $\sigma$ donation accumulates charge on the metal which donates back to the $\pi^*$ orbital of the carbonyl. An equilibrium is reached in which the (perhaps) neutral metal becomes bound to the neutral, 'poor $\sigma$ donor' carbonyl by both strong $\sigma$ and $\pi$ bonds. The $\sigma$-donor function of the carbonyl group is enhanced, or established, by the $\pi$ acidity of this ligand, and by the enhanced $\sigma$ acidity of the metal brought about by its ability to rid itself of excess charge through 'back-bonding'. The synergic back-bonding mechanism would fail if the metal were in too high an oxidation state for it would then lack the essential electron density to donate back to the carbonyl and so establish that ligand's $\sigma$-donor ability.

---

**Box 6-7**

Some prefer to introduce the back-bonding model by arguing first that $\sigma$ donation from the carbonyl ligand causes too great an accumulation of negative charge on the metal so that the metal then tends to establish its electroneutrality by back-donation to the carbonyl $\pi^*$ orbitals. Since the whole process is synergic, it matters little at what point in the cycle one begins the description. However, the present, perhaps unusual, path was chosen so as to highlight the initial poor $\sigma$ donor ability of the carbonyl ligand. Again, this is, no doubt, a matter of taste....

---

A similar account of the bonding in Zeise's salt, $K[PtCl_3(C_2H_4)]$, is offered by the Dewar-Chatt-Duncanson model. In this complex, the ethene ligand bonds 'side-on' to the metal. The synergic back-bonding, shown in Fig. 6-13, involves donation from the filled $\pi$ molecular orbital of the ethene together with back-bonding from the metal into the empty $\pi^*$ orbital of the ligand.

This bonding mechanism is expected for complexes of electron-rich metals with ligands offering both $\sigma$-donor and $\pi$-acceptor functions. The underlying driving force for synergic back-bonding derives from the operation of the electroneutrality principle together with the existence of two (or more) discrete electronic pathways to satisfy it. The Dewar-Chatt-Duncanson model is widely accepted. However, it provides us with a problem at this point. We have repeatedly emphasised the minimal roles of metal *d* orbitals in overlap with the ligands and yet the back-bonding mechanism can only succeed if the metal *d* orbitals overlap with the various ligand functions. It is curious that some argue that the (undoubted) validity of the Dewar-Chatt-Duncanson model demonstrates a universally active role for metal *d* functions in metal – ligand bonding orbitals. There is, however, another way.

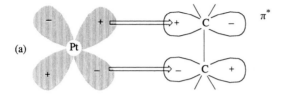

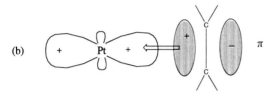

**Figure 6-13.** Synergic back-bonding in a platinum alkene complex. In (a), the interaction of a (filled) platinum $5d$ orbital with the $\pi^*$ molecular orbital of the alkene is shown, whilst in (b), the interaction of a $dsp$ hybrid orbital with the $\pi$ molecular orbital of the alkene is shown. Note that the two interactions result in electron density moving in opposite directions.

## 6.6  Valence Shells in High and Low Oxidation States

That way is to recognize that the nature of the valence shell is not constant but varies throughout the transition-metal series as, indeed, do most chemical priorities throughout the periodic table. The valence shell for Werner-type complexes, described in Section 2.2 in its limiting form, excludes the relatively tightly bound $3d$ orbitals. The exclusion is not total, of course, because of the tail of the $3d$ radial wavefunction and as evidenced by the (small) 'violation' of Laporte's rule. One expects the situation to be different in low oxidation state complexes like the carbonyls. Here, the lesser formal effective nuclear charge results in all electrons – $3d$, $4s$ and $4p$ orbitals – being less well bound, but to differing degrees. From our discussion of the electron configurations of transition-metal *atoms* in Chapter 1, for example, we learned that the $4s$ electrons are often more strongly bound than the $3d$ electrons. This arises, we recall, because of the more penetrating character of the $4s$ orbitals. They themselves are thus partly exposed to a higher effective nuclear charge and also their penetration serves to screen the $3d$ electrons more from the nucleus. Qualitatively, therefore, the differences between the radial forms of $3d$, $4s$ and $4p$ orbitals in higher and lower oxidation states are expected to follow the trends illustrated in Fig. 6.14. We see that, in low oxidation state species, the radial extent of the $3d$ orbitals is much more like that of the $4s$ and $4p$ orbitals than it is in higher oxidation state complexes. The valence shell now comprises all these orbital sets. Now the $d$ orbitals can overlap significantly with appropriate ligand functions and no conflict with the back-bonding model is evident. Furthermore, as the $d$ orbitals are now fully engaged in the bonding process and exposed to the envi-

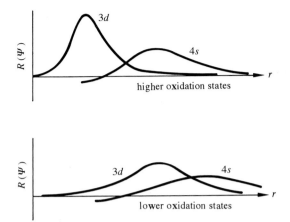

**Figure 6-14.** The 3*d* orbitals enter the valence shell in lower oxidation states.

ronment, one does not expect to find stable, open *d* shells. Unpaired electrons, for example, within the *d* or valence shells should be particularly unstable with respect to any process that fills the shells. In short, we expect to observe an *18-electron rule* [2 × (5 + 1 + 3) for the 3*d* + 4*s* + 4*p* subshells] governing electron counts, just as one sees the 8-electron rule within the first long series of the periodic table. Configurations with open shells and unpaired electrons in Werner-type complexes are stable and common (indeed, the norm) because the open *d* shells essentially lie inside the valence shell. As observed in the introductory survey of Chapter 1, there is no tendency towards organic-like free-radial behaviour with Werner-type complexes having unpaired electrons. With very few exceptions, the same is not true in carbonyl-type chemistry of transition metals in low oxidation states. Overall, therefore, we argue that the change from higher to lower oxidation state chemistry signals an important change of bond type in a way that the change from main group to Werner-type transition-metal chemistry does not.

This change in bond type, though discontinuous, is blurred. The passage from one oxidation state to another is discontinuous in the sense that it is associated with discrete additions or removals of individual electrons. It is blurred, on the other hand, because the electroneutrality principle will minimize any local charge concentrations. We saw one example of this in Section 6.1. The nephelauxetic effect arises because the radial distribution of the *d* shell expands as ligands donate negative charge into the more penetrating regions of the 4*s* and/or 4*p* shells. The nephelauxetic effect thus defines a spread of differential orbital expansion somewhat akin to the differences shown in Fig. 6.14. However, the range of radial variation from the nephelauxetic effect for any given formal oxidation state is, as commented at the end of Section 6.1, generally expected to be less than that characterizing different oxidation states. In Fig. 6-15, we provide a schematic indication of this 'blurring'.

Amongst the consequences to be expected from the change from Werner-type behaviour to carbonyl, low oxidation state chemistry is a breakdown in the efficacy

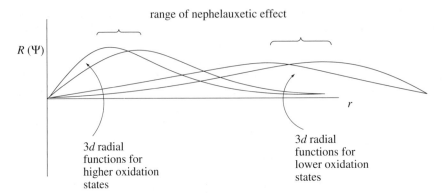

range of nephelauxetic effect

$R\,(\Psi)$

$r$

3d radial
functions for
higher oxidation
states

3d radial
functions for
lower oxidation
states

**Figure 6-15.** Nephelauxetic expansions of the *d* orbitals are expected to be less than those resulting from integral changes in oxidation state.

of the ligand-field method. Calculations of spectral splittings between states which are no longer well described by combinations of pure *d* functions cannot be expected to succeed well at all. As discussed in Section 6.4, ligand-field theory works within a domain – the ligand-field regime, if you like – and lower oxidation state complexes are expected to fall increasingly outside of that domain. Unfortunately, it may not be easy to demonstrate this expected failure experimentally. The greater orbital mixing and larger *d* orbital splittings expected for the low oxidation state type of valence shell generally result in decreased energies of charge-transfer bands and greater proximity of '*d–d*' transitions to these charge-transfer bands. The result is that spectral features that might be assigned as '*d–d*' type are frequently obscured by intense charge-transfer absorptions. A general paucity of apparent '*d–d*' transitions thus precludes any thorough testing of the gradual breakdown of the ligand-field method that is to be expected in low oxidation state transition-metal chemistry.

## 6.7  Electroneutrality and the Elasticity of the *d* Shell

The synergic, or interactive, nature of the back-bonding mechanism was strongly emphasized in Section 6.5. Despite that emphasis, some find it natural to focus upon the change on the ligand: $\pi$ back-donation by the metal enhances the $\sigma$ basicity of the carbonyl group. However, one might equally observe that carbon monoxide binds to transition metals because the metals act, in low oxidation states, as exceptionally good acceptors. Acceptors are atoms or molecules which readily absorb electron density. Transition metals can do this either because they start out as positively charged atoms (but not for carbonyls, of course, because these ligands are intrinsically poor $\sigma$ donors) or because they can divest themselves of the charge by passing it onto other atoms (ligands) (see Chapter 9) or, indeed, back to the original donor. There is even a third way in which transition metals can reduce charge concentration; namely, by the nephelauxetic effect. On receiving electron density from one or more ligands, the expansion of the *d* shell, which, as we have seen, may have little overlap with the ligand functions, spreads some of the metal electron density out over a larger volume of space. As expressed in Chapter 1, the electroneutrality principle asserts that atoms will acquire only small overall charge. However, the principle is more general than that, for it really means that charge will be more evenly distributed. The enlargement of an atom, or of a subset of its electron density, is an equally effective way of reducing charge density. The elasticity of approximately nonbonding *d* orbitals in transition metals confers upon the *d*-block metals an extraordinarily facile redox chemistry. Recall that an important subgroup of redox reactions in transition-metal chemistry is that involving stepwise gain or loss of electrons without bond rupture or overall geometry change. In part, this is to be laid at the door of elastic *d* orbitals. Of course, a great deal of redox chemistry does involve bond rupture and ligand change, thus also characterizing the broad spread of transition-metal chemistry, as it does elsewhere in the periodic table. Something of this is discussed in Chapter 9.

# 6.8  The Bonding Contributions of *d* Orbitals

We have emphasized the change of bond type that accompanies the growing participation of the metal *d* orbitals in the valence shell of transition-metal complexes on passing from high to low oxidation states. In order to make this point, we have perhaps overstressed the similarity between the Werner-type complexes in the *d* block and the chemistry of main group metals. We conclude this chapter, therefore, with the seeming *volte face* of asserting that, notwithstanding the small overlap of 3*d* orbitals with ligand orbitals in Werner-type compounds, their contribution to overall bonding is by no means negligible.

Once more, to make the point, let us take the extreme view that the 3*d* orbitals overlap negligibly with the ligand orbitals. Then, as described in Section 6.4, the repulsive or Coulombic interaction of the *d* electron density with that of the bonding electrons results in their differentiation as monitored by ligand-field splittings. As pointed out in Section 2.3 and explored in some detail in Chapter 7, that *inter*action impinges on the bonds themselves. The physical approach of metal and ligands may be frustrated to a greater or lesser extent by the *d* electron density. This hindrance will depend upon the spatial distribution of the *d* electron density and, in turn, upon whether the complex is in a 'high-spin' or 'low-spin' state, where that is appropriate. It is obvious then that the existence of the *d* electron density will generally have a marked effect upon net bond strengths and, on occasion, molecular geometry. Again, we shall see more of this in Chapter 7. In this sense, therefore, it would be appropriate to recognize that, even in higher oxidation state complexes, the *d* orbitals have a significant role in bonding even where their overlap with the ligands may be minimal.

Furthermore, as discussed in Section 6.7, the ability of the elastic *d* orbitals to function as electron 'sinks' contributes greatly to the rich variety of redox chemistry that is so characteristic of the *d*-block elements. Here too, therefore, we recognize the 'bonding' role of the *d* orbitals in Werner-type complexes as well as in carbonyl-type chemistry.

In many respects, this is the kernel of this book. For years it has not been too clear how one could consistently account for the wide variety of transition-metal chemistry in a way that does not conflict with the equally varied phenomena of spectroscopy and magnetochemistry that are so well rationalized by ligand-field theory. There is a tendency – psychologically quite natural, no doubt – for those interested in synthetic and mainstream chemistry not to look too closely at theory and physical properties, and, of course, *vice versa*. However, there has always been the need, surely, to build a logical synthesis of, or bridge between, these two aspects of the same subject. We hope that our presentation in this book goes some way towards providing that overview.

## Suggestions for further reading

1. M. Gerloch, *Ligand Field Parameters*, Cambridge University Press, Cambridge, **1973.**
2. C.K. Jørgensen, *Absorption Spectra and Chemical Bonding in Complexes*, Pergamon Press, Oxford, **1962.**
3. C.K. Jørgensen, *Modern Aspects of Ligand-Field Theory*, North Holland, **1970**.
   – These three references cover many aspects of the nephelauxetic and spectrochemical series.
4. J.N. Murrell, S.F.A. Kettle and J.M. Tedder, *Valence Theory,* 2nd ed., Wiley, New York, **1969.**
5. F.A. Cotton, *Chemical Applications of Group Theory,* 3rd ed., Wiley, New York, **1990.**
   – These two references are useful for molecular orbital theory of the water molecule.
6. M. Gerloch, R.G. Woolley, "The Functional Group in Ligand-Field Studies: The Empirical and Theoretical Status of the Angular Overlap Model", in *Prog. Inorg. Chem.*, **31**, 371.
7. M. Gerloch, "The Cellular Ligand-Field Model", in *Understanding Molecular Properties*, (Eds.: J.S. Avery, J.P. Dahl, A.E. Hansen), Reidel, **1987**, p.111.
   – Despite the title in reference (6), both references (6) and (7) describe the cellular ligand-field model.

# 7  Steric Effects of Open $d$ Shells

At the end of Chapter 2, we emphasized the *inter*action between the broadly non-bonding $d$ electrons and all other electrons housed in the bonding orbitals of Werner-type complexes. Most of the material we have covered since then has been concerned with the effects of the bonds upon the $d$ electrons. Now we turn to the effects of the $d$ electrons upon the bonds. We shall see that, although the $d$ orbitals overlap little with the ligand orbitals in Wernerian complexes, they do make significant contributions to what is collectively called the 'metal-ligand bonding'.

In a nutshell, the $d$ electrons repel the bonding electrons. They get in the way of the bonds and, to a greater or lesser degree, frustrate the attraction between metal and ligands. In essence, the proposed minimal overlap of $d$ orbitals with the ligands, but significant repulsive interaction with the bonds, is equivalent to a focus upon the two-electron operator rather than the one-electron operator; that is, upon repulsions rather than overlap.

## 7.1  Bond Lengths in Octahedral Complexes

Consider the repulsive effects of the $d$ electrons in a series of $ML_6$ complexes as the $d$ configuration of the central metal varies across the transition-metal series. All $d$ electron density will repel the bonding electron density. The effects on the $t_{2g}$ electron density will be relatively small, however, as these orbitals largely lie inbetween the bonding regions. On the other hand, $e_g$ electron density directly frustrates the bonding.

In Figure 7-1, we sketch the effects upon bond lengths predicted to arise from the repulsive role of the $d$ shell in both high-spin and low-spin octahedral species. Of course, experimental bond lengths are also expected to decrease across the period due to the usual increase in $Z_{eff}$ that arises from the monotonic increase in nuclear charge together with the imperfect self-shielding of non-core electrons. Figure 7-2 presents typical variations in ionic radii as determined from experimental metal–ligand bond length measurements for both divalent and trivalent metals of the first transition period. The qualitative agreement between theory and experiment is evident.

Two further consequences of the steric activity of open $d$ shells are also important. One, which might seem somewhat circular but does not, in fact, involve any 'double counting', is that longer bonds are accompanied by smaller ligand-fields, that is, by

**Box 7-1**

It is common to use plots of ionic radii, as in Fig. 7-2, for the transition metals as functions of $d^n$ configuration rather than the bond lengths. These are constructed by subtraction of appropriate ligand ionic radii from experimental bond lengths. As usual, one assumes an additivity between varying metal radii and a constant ligand radius. One might be concerned with two aspects of such a procedure: a) why use 'ionic' ligand radii rather than 'covalent', and b) might not the ligand 'radius' vary in response to the variable acidity of the metal across the series. We could avoid these imponderables by plotting typical bond lengths, noting that any variation in metal acceptor ability should vary with $Z_{eff}$ and the $d$ configuration as in Fig. 7-1, but it is not usually possible to obtain a suitable series of complexes with similar ligands.

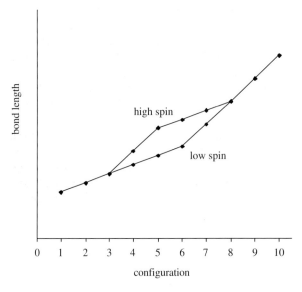

**Figure 7-1.** The repulsive effects of the $d$ shell on bond lengths. Small increases are expected with occupancy of the $t_{2g}$ subset, large ones with occupancy of the $e_g$.

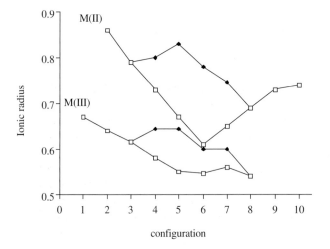

**Figure 7-2.** Effective ionic radii of high- and low-spin divalent and trivalent ions of the first row transition elements. The filled points represent high-spin ions.

smaller $Dq$ or $\Delta_{oct}$ values. The other is that variations in bond energies and the overall thermodynamic stabilities of complexes will accompany these bond length variations. We shall take up these themes in the next chapter.

## 7.2 Planar Coordination in $d^8$ Complexes

Planar coordinated systems, you will recall from Chapter 1, formed a major group of exceptions to the otherwise very successful geometry modelling of Kepert. That model explicitly neglected any steric role for the non bonding electrons, however. Let us now recognize and incorporate the steric activity of the $d$ shell in $d^8$ systems.

First, consider an octahedral nickel(II) complex. The strong-field ground configuration is $t_{2g}^6 e_g^2$. The repulsive interaction between the filled $t_{2g}$ subshell and the six octahedrally disposed bonds is cubically isotropic. That is to say, interactions between the $t_{2g}$ electrons and the bonding electrons are the same with respect to $x$, $y$ and $z$ directions. The same is true of the interactions between the six ligands and the exactly half-full $e_g$ subset. So, while the $d$ electrons in octahedrally coordinated nickel(II) complexes will repel all bonding electrons, no differentiation between bonds is to be expected. Octahedral $d^8$ coordination, *per se*, is stable in this regard.

Now consider a molecular stretching vibration that alternately elongates and compresses axial (parallel to $z$, let's say) and equatorial bonds as outlined in Fig. 7-3. Imagine an extreme vibration of this kind that eventually distorts an octahedral molecule so as to gradually remove two *trans* ligands (again, let this direction be

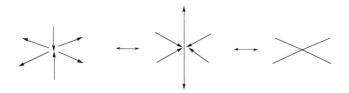

**Figure 7-3.** Vibrational distortion leading to a planar complex.

taken as $z$). The ligand-field splitting diagram changes as illustrated in Fig. 7-4. For ligands regarded as point negative charges *or* as $\sigma$ donors, the ligand-field along $z$ decreases with respect to that along $x$ or $y$. A simple mnemonic is that "elongation along $z$ stabilizes orbitals with the letter $z$ in them"; $d_{z^2}$, is less repelled and more stabilized than $d_{x^2-y^2}$ while $d_{xz}$ and $d_{yz}$ are more stable than $d_{xy}$, although the splitting of the $t_{2g}$ orbital set is less than that of the $e_g$ set because the $t_{2g}$ orbitals are less closely directed towards the ligands. As shown in Fig. 7-4, a sufficiently large axial elongation of the octahedron, accompanied by a commensurate shortening of the equatorial bonds in response to the electroneutrality principle, raises the energy of the $d_{x^2-y^2}$ sufficiently that the electrons pair up in the $d_{z^2}$ orbital, despite the penalty

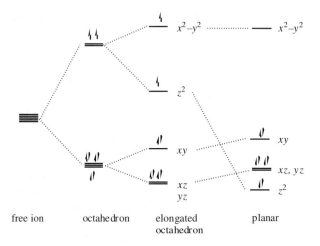

free ion        octahedron        elongated        planar
                                   octahedron

**Figure 7-4.** Schematic splitting of *d* orbitals in a planar environment. The ordering of the lowest four orbitals in unimportant here. The $x^2 - y^2$ orbital lies much higher in energy than the other *d* orbitals, and the low-spin arrangement follows.

of pairing energy. In short, the molecule adopts a low-spin $(xz, \, yz)^4(xy)^2(z^2)^2$ configuration rather than the high-spin $(xz, \, yz, \, xy)^6(z^2)^1(x^2 - y^2)^1$ configuration of the near-octahedron. Now note that the doubly filled $d_{z^2}$ orbital in the low-spin configuration offers two electrons' worth of repulsion along the *z* direction while the empty $d_{x^2-y^2}$ orbital provides no repulsion at all to the equatorial ligands. In other words, the planar coordination, which we imagined as being achieved by drawing out two *trans* ligands from a perfectly stable octahedron, is seen to be stable also with respect to *d*-electron repulsions. Furthermore, the $d_{z^2}$ electron pair situated above and below the coordination plane provides a strong disincentive to the return of the axial ligands; that is to say, this nonbonding lone pair of electrons tends to frustrate donor addition to, or adduct formation with, planar nickel(II) species. Also completing what at root is a cyclic, effectively synergic, process, we note that the absence of *d*-electron repulsions from the $d_{x^2-y^2}$ orbital encourages the closer approach of the four ligands. This in turn allows these four ligands to satisfy the acidity of the metal atom as the six more distant ligands did in the octahedron. Altogether, therefore, we see that both octahedral and planar geometries are stable with respect to the steric activity of the open *d* shell.

We also see how a planar geometry for $d^8$ complexes can be preferred over a tetrahedral one. With no regard to the steric role of the *d* shell, one expects, with Kepert, to observe tetrahedral geometry for all four-coordinate complexes. On the other hand, should planar coordination be once achieved for the $d^8$ configuration, it will resist distortion towards the tetrahedron because of the repulsive effect of the lone pair normal to the plane. Kepert has observed that the placing of even one or two tenths of an electron charge between tetrahedrally disposed ligands on each side of, and close to, the metal directly favours a switch to planar geometry. In other words, four ligands plus two half lone pairs adopt an octahedral array. In short, Kepert's basic model needs to be supplemented by a recognition of the steric role of an open $d$ shell.

If this is so, then how is it that his model works so well in other cases without that addition? Well in many, though not all, cases, the additional effects of the $d$-electron repulsions are to modify bond lengths rather than bond angles. We discuss such an example in the next section. Before doing so, however, there is more to say about planar coordination.

We have argued that, once achieved, planar coordination in $d^8$ systems is stable with respect to higher coordination number or tetrahedral distortion. The question arises then about what circumstances favour planarity in the first place. In particular, we enquire about the occurrence of tetrahedral verses square planar stereochemistry for $d^8$ complexes. Why, for example, is the $[Ni(CN)_4]^{2-}$ ion planar but $[NiCl_4]^{2-}$ tetrahedral?

First, note that there is a parallel relationship between high-spin tetrahedral $d^8$ and spin-paired planar $d^8$, as compared with the octahedral and planar situations just described. Analogous to Fig. 7-4, we have Fig. 7-5. Do not be confused about the reversed labelling of the $xy$ and $x^2-y^2$ orbitals at the extremes of Fig. 7-4 and 7-5 for the plane. The reversal is an artifact of the way we define the global axis frames for the tetrahedron and octahedron (see Figs. 3-2 and 3-6). Thus, on squashing a tetrahedron to a square plane, we find the M−L bonds lying *inbetween* the $x$ and $y$ axes while they lie *along* these axes for the situation depicted in Fig. 7-4.

Once again we see how the planar geometry is stabilized by removal of 'repulsive' electrons from the $d_{xy}$ (ligand-directed) orbital. The achievement of planar

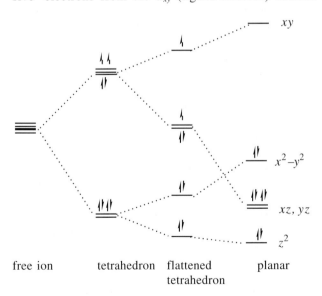

**Figure 7-5.** Compare with Fig. 7-4, but note the change of axis frame so that $xy$(oct) → $x^2-y^2$(tet).

coordination again involves the pairing up of electrons and the contest between the ligand-field promotion energy ($E(xy) - E(x^2-y^2)$) and the pairing energy. That contest is most likely to be resolved in favour of the planar, low-spin arrangement for ligands higher in the spectrochemical series giving rise to greater ligand-field splittings. Thus, we find planar $[Ni(CN)_4]^{2-}$ but tetrahedral $[NiCl_4]^{2-}$.

---

**Box 7-2**

In principle, we might expect to observe the low-spin $d^8$ arrangement even in a distorted tetrahedral complex, provided that the associated ligand-field splittings are large enough. One such example is found in the complex $[NiI_2(Ph_2PCH_2CH_2OCH_2CH_2OCH_2CH_2PPh_2)]$:

---

The change from high- to low-spin $d^8$ configurations is necessarily discontinuous. A given complex is either on one side of the divide or the other. We conclude this section with a look at how the steric role of the $d$ shell can affect angular geometries within a series of just high-spin, nominally tetrahedral nickel(II) complexes.

Before $[Et_4N]_2[NiCl_4]$, containing nearly tetrahedral $[NiCl_4]^{2-}$ ions was first synthesized, it was thought that 'tetrahedral' geometry must be forced in four-coordinate nickel(II) species by including into the coordination shell such bulky groups as triphenylphosphine. Indeed, the very first non-planar, four-coordinate nickel(II) complexes to be prepared were the bis-halo-bis(triphenylphosphine) nickel(II) molecules, $NiX_2(PPh_3)_2$ X = Cl, Br, I. It was no surprise to find that none of these molecules possessed very near tetrahedral symmetry. It was puzzling, however, to observe that their geometries deviate from tetrahedral towards planar coordination *increasingly* along the series X = Cl, Br, I. In fact, the bis-iodo complex is planar coordinated. The increasing bulk of the halogens along this series might have led one to expect deviations from planarity that increase as iodine replaces bromine, or as bromine replaces chlorine. Clearly, the increased flattening of the tetrahedron on passing from the bis-chloro through to the bis-iodo complex occurs in spite of, rather than because of, steric repulsion between the ligands.

Recall, then, the positions of the halogens within the nephelauxetic series. More negative charge is donated to the central metal from the iodine ligands than from the bromine ligands, which in turn donate more than the chlorine ligands. Furthermore, it has been shown by modern ligand-field analysis that the extent of these charge donations is greater in the bis-phosphine complexes than in the tetra-halo complexes as a result of the $\pi$ acidity of the phosphine ligands. Refer now to the splitting diagram for flattened tetrahedral $d^8$ species in Fig. 7-5. While the four $t_2$ electrons in the pure tetrahedron are distributed equally amongst the three $d$ orbitals, they are now arranged in favour of the $d_{xz}$, $d_{yz}$ orbital pair. Conversely, there is less $d$ electron density in the $d_{xy}$ orbital – the '$t_2$' orbital most nearly directed at the ligands in the flattened tetrahedral coordination. Overall, the tendency of a tetrahedron to distort towards a plane will be greater the more important the repulsive interactions between the $d$ electrons and the bond orbitals become. Those repulsions are expected to increase as the $d$ orbitals expand in response to the increasing nephelauxetic effect along the series Cl < Br < I. In short, we argue that it is the $d$-electron – bonding electron repulsions that determine the relative angular geometries of these bis-halo-bis-phosphine nickel(II) complexes.

## 7.3  Trigonal Bipyramidal Coordination

The steric activity of open $d$ shells is well illustrated by the ligand fields and bond lengths in five-coordinate complexes with formal trigonal bipyramidal geometry. Consider the series of complexes [M$^{II}$(Me$_6$tren)Br]$^+$, where Me$_6$tren is tris((dimethyl-amino)ethyl)amine. The Me$_6$tren ligand is a tripodal tetraamine donor and the coordination geometry of this series of complexes is shown in Fig. 7-6. Metal –

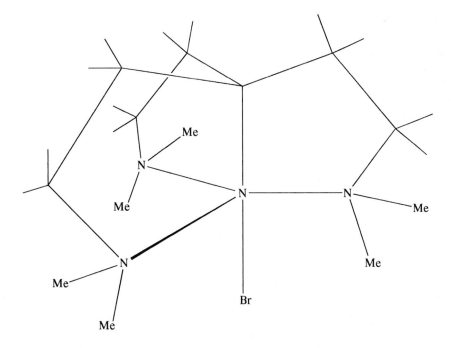

**Figure 7-6.** Coordination geometry in [M$^{II}$(Me$_6$tren)Br]$^+$ species.

ligand bond lengths for these molecules with M = Co(II), Ni(II), Cu(II) and Zn(II) are listed in Table 7-1. Our discussion focuses upon the differences between the axial and equatorial metal–nitrogen distances.

**Table 7-1.** Bond lengths and $d$ orbital configurations for $[M^{II}(Me_6tren)Br]^+$ cations.

|  | Co(II) | Ni(II) | Cu(II) | Zn(II) |
|---|---|---|---|---|
| M–N(ax)/ Å | 2.15 | 2.10 | 2.07 | 2.19 |
| M–N(eq)/ Å | 2.08 | 2.13 | 2.14 | 2.11 |
| M–Br/ Å | 2.43 | 2.47 | 2.39 | 2.45 |
| $z^2$ | • | • | • | •• |
| $xy,\ x^2-y^2$ | : | :. | :: | :: |
| $xz,\ yz$ | :: | :: | :: | :: |

Consider first what pattern would be expected for the zinc(II) complex. Here, the $d$ shell is full ($d^{10}$) and therefore offers equal repulsion in all directions. As such, we should be able to make a prediction based solely upon bond-pair repulsions as in VSEPR theory. We predict that the axial Zn–N bond should be longer than the equatorial Zn–N bonds because it suffers three bond-bond repulsions at 90° while each equatorial bond suffers only two such repulsions at 90°. The experimental bond lengths in Table 7-1 support this view. Now consider the cobalt(II) complex. In the three-fold symmetry of the trigonal bipyramid, the $d$ orbitals split up as shown in Table 7-1. The $d_{z^2}$ orbital is highest in energy because it points directly at the ligands, the degenerate $d_{xy}$, $d_{x^2-y^2}$ pair lies in the plane of the equatorial ligands and the electrons in these two orbitals are the next most repelled, and the degenerate $d_{xz}$, $d_{yz}$ pair is directed inbetween the ligands and is the most stable. All complexes in this series are high-spin. Their strong-field configurations are indicated in Table 7-1. In each case, the $d_{xz}$, $d_{yz}$ orbital subshell is full and, for the cobalt complex, there is one electron in each of the remaining three $d$ orbitals (in effect, one in the $d_{z^2}$ orbital directed mostly along $z$ and two in the $d_{xy}$, $d_{xz^2-y^2}$ pair involved, between them, with the $x$ and $y$ directions). The repulsions offered towards the axial and equatorial ligands by this electronic arrangement are thus essentially equal. We therefore expect to find the same relationship between axial and equatorial bond lengths in this cobalt(II) complex as in the zinc(II). The bond lengths in Table 7-1 support this. Although their absolute values differ between the $d^7$ and $d^{10}$ complexes, of course, because of the overall trend in $Z_{eff}$ to increase across any period, we do find longer axial and shorter equatorial Co–N bonds as predicted. Turning now to the $d^9$ copper(II) complex, we note that the $d_{xy}$, $d_{x^2-y^2}$ orbital pair is full while the $d_{z^2}$ orbital still houses only one electron. The $d$-orbital to ligand orbital repulsions in the equatorial plane are thus much greater than along the three-fold axis and the bond length pattern is reversed. What is actually observed is a marked shortening of the axial bond. This results from the combined effects of the change in steric activity of the open $d$ shell and of the increased effective nuclear charge on replacing cobalt by copper. Copper is the stronger acceptor and the drive towards electroneutrality is satisfied most easily by a close approach of the axial ligands since they are least hindered by the $d$-electron distribution.

These bond length variations are complemented by analogous changes in the strengths of the local ligand fields associated with the various ligands. Local ligand-field strengths are represented and monitored in a version of the ligand-field model called *cellular ligand-field theory*. While space and level do not allow any full discussion of this powerful modern approach, some idea of its basis is presented in Box 7-3 for those with an interest.

# 7.4  The Jahn-Teller Effect

A somewhat abstruse group-theoretical (symmetry based) theorem was published in the late 1930's by Jahn and Teller. It is in effect that "For non-linear molecules, a nuclear configuration which begets an orbitally degenerate occupied state is unstable with respect to one without such orbital degeneracy." In the 1950's, Orgel exploited this theorem to rationalize anomalous geometrical features of copper(II) and chromium(II) compounds. Empirically, it is found that formally octahedral or tetrahedral $d^9$ complexes are highly distorted and, to a slightly lesser extent, the same is true of high-spin octahedral $d^4$ species also. Typical of these distortions is that while equatorial metal–oxygen bond lengths, for example, in 'octahedral' copper complexes are about 2.0 Å, axial bonds for two *trans* ligations take values anywhere between 2.3 and 2.9 Å. These are large effects and are not observed in complexes of metal ions with other $d^n$ configurations, unless caused by apparent ligand constraints (chelation, for example). The explanation of this effect – generally known as the Jahn-Teller effect – exploits the Jahn-Teller theorem.

In octahedral symmetry, the ground term of the $d^9$ configuration is $^2E_g$, as discussed in Chapter 3. This is an orbitally degenerate state and hence subject to some nuclear distortion that removes the degeneracy. The situation is represented in Fig. 7-7 where we investigate the effects of a tetragonal molecular distortion. Since the Jahn-Teller theorem does not determine the type of distortion that must occur, we look at the tetragonal one simply because most 'octahedral' copper(II) distortions roughly conform to this. In Fig. 7-7, we see that the $e_g^3$ configuration of the regular octahedron is degenerate ($^2E_g$) because the hole may be sited in the $d_{z^2}$ or $d_{x^2-y^2}$ orbitals with equal probability. Tetragonal elongation **or** compression of the octa-hedron removes the degeneracy of the $e_g$ orbital pair to leave $^2A_{1g}$ or $^2B_{1g}$ – corresponding to the hole being housed in the $d_{z^2}$ or $d_{x^2-y^2}$ orbital respectively – ground terms which are, of course, nondegenerate. *The driving force for the distortion is, once again, the steric activity of the open d shell.*

In high-spin species (to which the theorem is *not* restricted), perusal of the appropriate configurations for octahedral complexes across the transition period (see Fig. 5-1) shows that $d^4$ and $d^9$ configurations are candidates for the Jahn-Teller

**Box 7-3**

Consider the local interactions between *d* orbitals, referred to a local frame, and various bond orbitals. In Fig. A, we represent such interactions for orbitals characterizing local σ

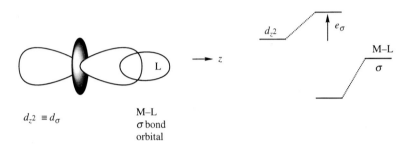

Figure A

symmetry. The $d_\sigma$ orbital ($d_{z^2}$) suffers a shift in energy that we label $e_\sigma$ (we use the lower-case *e* for the energy shift of an *orbital*.). The equivalent situation for a π interaction in the local *xz* plane is shown in Fig. B and defines the local ligand-field parameter $e_{\pi x}$. An analogous

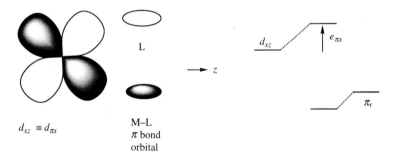

Figure B

interaction in the *yz* plane defines $e_{\pi y}$. These various parameters relate directly to the nature of any σ, $\pi_x$ or $\pi_y$ bonding within the local region of space. Thus, for a metal-pyridine ligation, for example, one expects that bonding orbitals between metal (4*s*/4*p*) orbitals and ligand orbitals will exist for σ and $\pi_\perp$ (where ⊥ means perpendicular to the pyridine plane) interactions but not for $\pi_{//}$. Detailed analysis confirms this sort of prediction. The (global) ligand field for the complex as a whole is constructed by appropriate additions of such local, or cellular, ligand fields. This powerful approach provides a means by which analysis of global ligand-field phenomena – spectral transition energies, intensities, optical activity as well as various magnetic properties – may probe the underlying bonding in a complex.

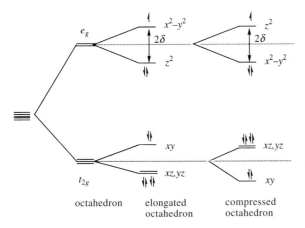

**Figure 7-7.** Splitting of $d$ orbitals for elongated and compressed octahedral fields.

---

**Box 7-4**

A tetragonal distortion is one that maintains the four-fold symmetry of the octahedron. Here, we consider two equal *trans* metal–ligand bond lengths being different from the other (equal) four. A trigonal distortion maintains the three-fold symmetry from the octahedron. This might be effected by six equal bond lengths with unequal interbond angles:

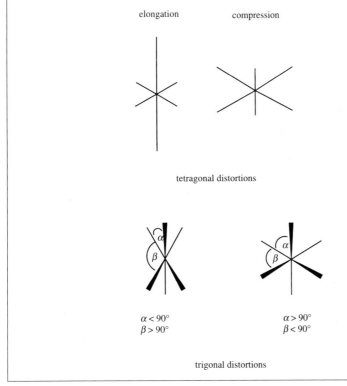

effect because of orbital degeneracy in the $e_g$ configurations and $d^1$, $d^2$, $d^6$, and $d^7$ configurations because of orbital degeneracy in their $t_{2g}$ shells. However, Jahn-Teller effects arising out of $t_{2g}$ shells that are incompletely (but not exactly half-) filled are much smaller in magnitude than those associated with $e_g^1$ or $e_g^3$ configurations. This is because the $t_{2g}$ orbitals interact less with the bonds than do the $e_g$ orbitals, being directed inbetween them. Once again, in evidence of the *inter*active nature of $d$ electrons in the ligand environment, the splitting of the $t_{2g}$ orbitals due to Jahn-Teller distortion is small and the distortion due to the unevenly filled $t_{2g}$ subshell is also small. We comment further on the difference between unevenly filled $t_{2g}$ and $e_g$ shells shortly. For octahedra, only $d^4$ and $d^9$ configurations are expected – and observed – to suffer significant Jahn-Teller distortions.

For tetrahedra, orbital degeneracies in the higher-lying $t_2$ orbital subset should give rise to larger distortions than in the $e$ set. On this basis, one expects distorted tetrahedra for $d^3$, $d^4$, $d^8$ and $d^9$ configurations. Tetrahedral $d^3$ and $d^4$ complexes are very rare. Tetrahedral $d^8$ complexes are reasonably common but occur often with reasonably regular geometries. It is likely that the lack of any significant Jahn-Teller distortion here is due to the strong-field limit being a poor description of the many-electron ground state in these systems. Tetrahedral copper(II), $d^9$, complexes, on the other hand, show large distortions, most frequently in the form of a flattening towards square planar geometry (see Box 7-5).

Returning to the octahedral species, there is one curious feature to the types of distortion observed in practice. The scheme in Fig. 7-8 shows how the distorted octahedron acquires a stability over the regular octahedron in $d^4$ and $d^9$ systems equal to one half of the splitting of the $e_g$ orbital pair. This arises as follows. To maintain the same mean ligand-field strength – in effect, to satisfy the electroneutrality principle to the same extent – a lengthening of two *trans* bond lengths is accompanied by a (lesser) shortening of the four equatorial bonds, or vice versa: this results in a barycentre splitting of the $e_g$ orbitals as shown in Fig. 7-7. If we label that splitting as $2\delta$ with respect to the energy of the octahedral $e_g$ orbitals,

---

**Box 7-5**

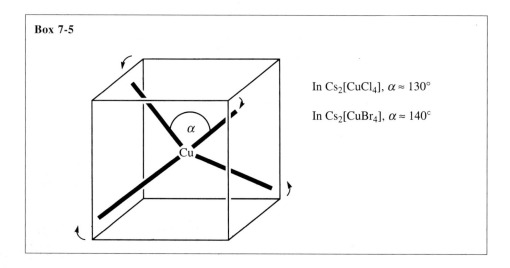

In $Cs_2[CuCl_4]$, $\alpha \approx 130°$

In $Cs_2[CuBr_4]$, $\alpha \approx 140°$

the $d_{z^2}$ is stabilized by $\delta$ upon elongation, for example, while the $d_{x^2-y^2}$ is destabilized by the same amount. The lower energy orbital houses two electrons (stabilization $2\delta$) while the higher one houses just one electron (destabilization $\delta$), giving an overall stabilization of $\delta$ which is one half of the $e_g$ orbital splitting due to distortion. No stabilization is associated with the splitting of the $t_{2g}$ orbital set here because this set is full in $d^9$ or exactly half-filled in $d^4$. Since the foregoing argument is symmetrical with respect to elongation or compression of the octahedron, the labels on the two members of the $e_g$ set are irrelevant. We might therefore expect elongated octahedra to occur roughly as often as compressed ones. Empirically, however, very many more elongated chromium(II) complexes are observed than compressed ones, and no compressed copper(II) complexes are observed at all (ignoring any with that geometry imposed by chelate ring strain and the like). Clearly, our explanation is incomplete.

Crystal-field theory accounts for $d$-orbital energy shifts in terms of the differential repulsive effects of negative point charges: the various $d$ orbitals are *raised* in energy. Ligand-field theory also refers to a raising of $d$-orbital energies, at least when the bond orbitals with which they interact are lower in energy than the $d$ orbitals: such is the case for ligands acting in $\sigma$- or $\pi$-*donor* modes. Occasionally, $d$-orbital energy shifts can be to *lower* energies, however. One case with which we are familiar is when a bonding orbital is higher in energy than the $d$ orbital, as for ligations in the $\pi$-acceptor mode (see Section 6.3.6). Another case, not discussed so far, is when a $d$ orbital interacts significantly with another suitable orbital of higher energy. Here we consider the antibonding $a_{1g}$ orbital of predominantly $4s$ parentage (see Fig. 7-8).

In strict octahedral symmetry, no proximate bonding or antibonding orbital arising from outside the $d$ shell has the same symmetry as the members of the $d$ shell. In the tetragonal symmetry of the present distorted octahedra, however, this is no longer true. The $d$-orbital symmetry labels in the tetragonal ($D_{4h}$ point symmetry group) environment are: $d_{z^2}$ ($a_{1g}$); $d_{x^2-y^2}$ ($b_{1g}$); $d_{xy}$($b_{2g}$); $d_{xz}d_{yz}$($e_g$). The predominantly $4s$ antibonding orbital, labelled $a_{1g}$ in octahedral symmetry, is still of $a_{1g}$ symmetry in the tetragonal environment. There arises, therefore, the possibility of interaction between, in effect, the metal $4s$ orbital and the $d_{z^2}$ orbital, but *only* the $d_{z^2}$ orbital. As the energy of the $4s$ orbital is higher than that of the $d_{z^2}$, the interaction is such as to stabilize the $d_{z^2}$ orbital further. This is true for both elongated and compressed octahedral geometries as shown in Fig. 7-8. We can expect the extra stabilization, $\delta'$ of the $d_{z^2}$ orbital to be roughly similar for either sense of distortion. However, in the elongated octahedron, two electrons occupy the $d_{z^2}$ but only one in the compressed geometry. This provides the asymmetry we seek. The energy gap between the various $d$ functions and the antibonding $a_{1g}^*$ is expected to be much less at the right end of a transition period than the left because of the more penetrating character of the $4s$ metal orbital together with the increased effective nuclear charge at the right end. So this extra stabilization, due to $d$–$s$ interactions, that favours elongated geometries over compressed is expected to be more significant for copper(II) complexes than for chromium(II). As noted above, both geometry types are possibly observed experimentally for the $d^4$ system, but only the elongated one for $d^9$.

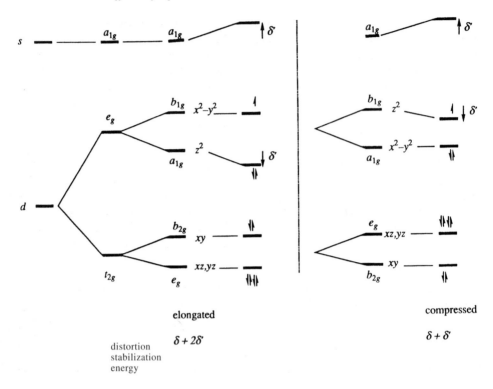

**Figure 7-8.** Orbital splitting diagrams for $d^9$ complexes with elongated and compressed octahedra.

The driving force for Jahn-Teller distortions in transition-metal complexes is the open $d$ shell. It is likely that explanations for them along the lines given above would have come about even if the *theorem* of Jahn and Teller had not been discovered. We make this remark not to denigrate that powerful piece of work, but as an attempt to defuse any mystery that might otherwise attach to Orgel's application of that group-theoretical construction.

The smaller magnitudes of distortions associated with unevenly filled $t_{2g}$ subshells in octahedral complexes relative to those deriving from unevenly filled $e_g$ subshells invites a little thought. From Fig. 7-7 we see how a distorted geometry for an open $e_g$ shell is more stable than an undistorted one. The splitting, $2\delta$, is a measure of that relative stability. Suppose for the moment that one were to focus upon that splitting energy as an absolute goal that is to be achieved in various situations. That is to say, suppose the measure of the relative stability of the distorted over the undistorted geometry is just given by the low-symmetry field splitting. If so, we might expect a larger physical distortion for open $t_{2g}$ shells than for open $e_g$ shells. That is because, for a given distortion, the splitting of the $e_g$ orbitals which point directly at the ligands is greater than that of the $t_{2g}$ orbitals directed inbetween. This predicts a result which is contrary to empirical fact. If, on the other hand, we recognize the driving force for the distortion as the steric activity of the open $d$ shell, we see immediately that the repulsive hindrance of $t_{2g}$ electron density is less than that of the ligand-directed $e_g$ density. An octahedral molecule with an open $e_g$ shell will distort more than one with an open $t_{2g}$ shell. In *consequence*, certain orbital splittings occur and *confirm*, as it were, the distortion in the manner of Fig. 7-7. However, there is no simple or direct way in which we can estimate the magnitudes of these orbital splittings and their associated stabilizing effects *a priori*.

The Jahn-Teller effect is pervasive. We have described its manifestation in electronic ground states leading to static distortion: the so-called '*static Jahn-Teller effect*'. When the tendency to distort involves smaller energies that are comparable with either spin-orbit coupling or vibrational energies, static distortions may not be observed. Instead, strong coupling of electronic and nuclear motions may result and give rise to the '*dynamic Jahn-Teller effect*'. Unfortunately we cannot pursue this matter here. We further note that 'unexpected' splittings are sometimes observed in '$d$–$d$' spectral bands. These have been ascribed to Jahn-Teller effects in excited states.

---

**Box 7-6**

*Example:* The spectrum of $[Ti(H_2O)_6]^{3+}$ ions, whose ground state geometry is nearly perfectly octahedral, is characterized by a large splitting.

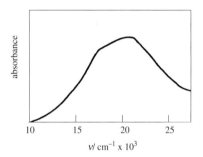

Only one band maximum is expected, of course, corresponding to the $^2T_{2g} \rightarrow {}^2E_g$ transition. The splitting of the asymmetric peak is ascribed to a Jahn-Teller splitting of the excited state which latter involves the open $e_g$ configuration $e_g^1$.

---

Finally, we should recognize that there can also be second-order Jahn-Teller effects. Above, we address orbital degeneracies within the ground state. However, since spin-orbit coupling, for example, can admix orbitally degenerate excited states into an orbitally non-degenerate ground state, Jahn-Teller effects can be observed in second-order. Once again, we do no more than mention these matters here.

# Suggestions for further reading

1. D.L. Kepert, *Inorganic Stereochemistry*, Springer-Verlag, Berlin, **1982**.
   – The scope of Kepert's model is, herein, made plain.
2. J.S. Griffith, *Theory of Transition Metal Ions,* Cambridge University Press, Cambridge, **1961**.
3. C.J. Ballhausen, *Molecular Electronic Structures of Transition Metal Complexes,* McGraw-Hill, New York, **1979.**
   – This last reference is for those who would like to see the Jahn-Teller theorem at a technical level.

# 8 Complex Stability and Energetics

## 8.1 The Thermodynamic Stability of Complexes

The thermodynamic stability of coordination compounds is relatively easy to determine, and provides us with a valuable pool of data from which we may assess the importance of ligand-field and other effects upon the overall properties of transition-metal compounds. The bulk of this chapter will be concerned with the thermodynamic stability of transition-metal compounds, but we will briefly consider kinetic factors at the close.

The stability of a complex is conveniently expressed in terms of the thermodynamic stepwise stability constant $K$ as defined in Eq. (8.1).

$$ML_{(n-1)} + L \rightleftharpoons ML_n$$
$$K = [ML_n] / ([ML_{(n-1)}][L]) \tag{8.1}$$

We should note at this point, that the above reaction implicitly refers to aqueous solutions, and that, for convenience, we have explicitly excluded free and coordinated solvent molecules. Strictly, the above relationships should be written as in Eq. (8.2).

$$M(H_2O)_{(7-n)}L_{(n-1)} + L \rightleftharpoons M(H_2O)_{6-n}L_n + H_2O$$
$$K = [M(H_2O)_{6-n}L_n][H_2O] / ([M(H_2O)_{(7-n)}L_{(n-1)}][L]) \tag{8.2}$$

For obvious reasons, we tend to use the simpler form, although we will discuss some of the limitations shortly. We may also consider overall stability constants, $\beta_n$ (Eq. 8.3).

$$M + nL \rightleftharpoons ML_n$$
$$\beta_n = [ML_n] / [M][L]^n \tag{8.3}$$

There is an obvious relationship between $K_n$ and $\beta_n$ as expressed in Eq. (8.4).

$$\log_{10} \beta_n = \sum_0^n \log_{10} K_n \tag{8.4}$$

Transition-metal complexes span an enormous range of stabilities. One of the principal aims of this chapter is to attempt to understand some of the factors which control these, and to determine the importance of ligand-field effects. Very extensive compilations of stability constants are available.

## 8.2  The Chelate Effect and Polydentate Ligands

Ligands containing more than one donor atom which can bond to a metal centre are termed *polydentate* or *multidentate*. Such ligands are extremely important and have played crucial roles in the development of coordination chemistry. Well known examples include 1,2-diaminoethane (ethylenediamine, *en*); 2,4-pentanedionate (acetylacetonate, *acac⁻*); 2,2'-bipyridine (*bpy*); and 1,2-diaminoethane-*N, N, N', N'*-tetraacetate (*edta⁴⁻*) (Fig. 8-1).

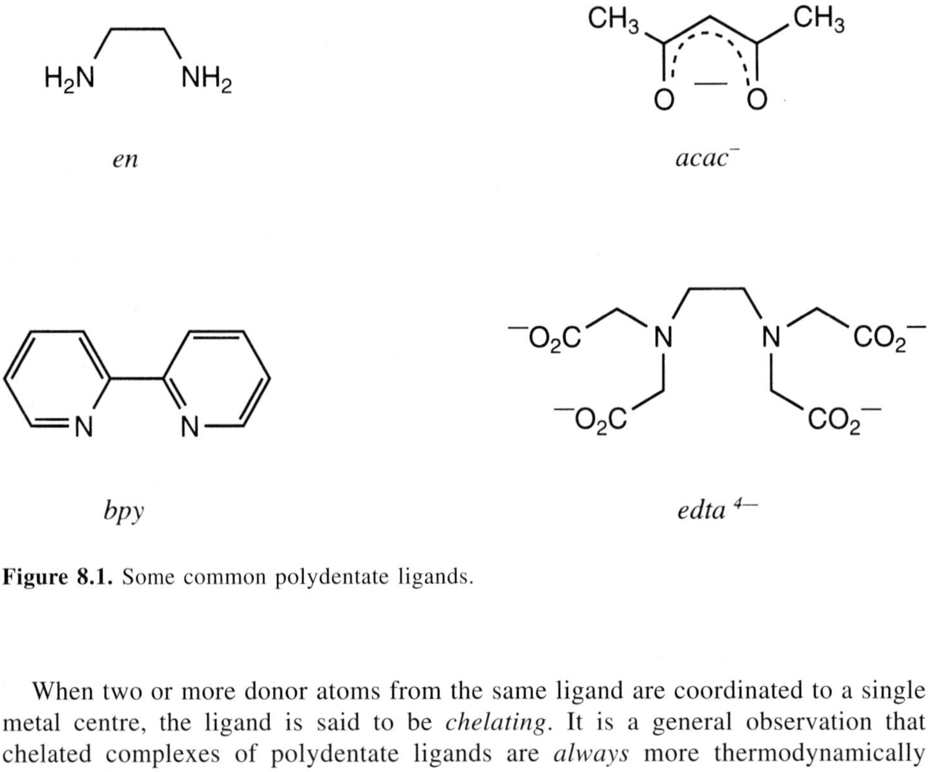

**Figure 8.1.** Some common polydentate ligands.

When two or more donor atoms from the same ligand are coordinated to a single metal centre, the ligand is said to be *chelating*. It is a general observation that chelated complexes of polydentate ligands are *always* more thermodynamically stable than those of the same metal with an equivalent number of comparable monodentate ligands. That is to say, the equilibrium

$$ML_n + (LL)_{n/2} \rightleftharpoons M(LL)_{n/2} + nL$$
$$(L = \text{monodentate ligand, LL} = \text{didentate ligand})$$

lies to the right. This is exemplified in the data for the $Ni^{2+}$/ *en*/ $NH_3$ system presented in Table 8-1.

**Table 8-1.** Stability constants for some nickel(II) complexes of ammonia and 1,2-diaminoethane.

|  | $log_{10} K$ |
|---|---|
| $Ni^{2+} + 2NH_3 \rightleftharpoons Ni(NH_3)_2^{2+}$ | 5.04 |
| $Ni(NH_3)_2^{2+} + 2NH_3 \rightleftharpoons Ni(NH_3)_4^{2+}$ | 2.92 |
| $Ni(NH_3)_4^{2+} + 2NH_3 \rightleftharpoons Ni(NH_3)_6^{2+}$ | 0.78 |
| $Ni^{2+} + en \rightleftharpoons Ni(en)^{2+}$ | 7.45 |
| $Ni(en)^{2+} + en \rightleftharpoons Ni(en)_2^{2+}$ | 6.23 |
| $Ni(en)_2^{2+} + en \rightleftharpoons Ni(en)_3^{2+}$ | 4.34 |
| $Ni(NH_3)_2^{2+} + en \rightleftharpoons Ni(en)^{2+} + 2NH_3$ | 2.41 |
| $Ni(NH_3)_4^{2+} + 2en \rightleftharpoons Ni(en)_2^{2+} + 4NH_3$ | 3.31 |
| $Ni(NH_3)_6^{2+} + 3en \rightleftharpoons Ni(en)_3^{2+} + 6NH_3$ | 3.56 |

## 8.2.1  Thermodynamic Origins of the Chelate Effect

We may relate the equilibrium constant to the free energy change for a reaction (Eqs. 8.5 – 8.7).

$$\Delta G^{\ominus} = - RT\ln K \qquad (8.5)$$
$$\Delta G^{\ominus} = \Delta H - T\Delta S \qquad (8.6)$$
$$\ln K = (\Delta S/R) - (\Delta H/RT) \qquad (8.7)$$

The various thermodynamic terms for the formation of 1:1 complexes of transition metals with 1,2-diaminoethane are presented in Table 8-2.

**Table 8-2.** Thermodynamic terms for the formation of 1:1 complexes of first row transition metals with 1,2-diaminoethane.

| $M^{2+}$ | $\Delta G$/kJ mol$^{-1}$ | $\Delta H$/kJ mol$^{-1}$ | $T\Delta S$ (298K)/kJ mol$^{-1}$ |
|---|---|---|---|
| Mn | $-15.9$ | $-11.7$ | 4.2 |
| Fe | $-24.7$ | $-21.3$ | 3.3 |
| Co | $-33.9$ | $-28.8$ | 5.0 |
| Ni | $-43.9$ | $-37.2$ | 6.7 |
| Cu | $-61.0$ | $-54.4$ | 6.6 |
| Zn | $-33.0$ | $-28.0$ | 5.0 |

$$M(H_2O)_6^{2+} + en \rightleftharpoons M(en)(H_2O)_4^{2+} + 2H_2O$$

In each case, both the entropy and enthalpy terms favour the formation of the chelated complex, regardless of the $d$-electron configuration. Note, however, that outside the $d$ block, i.e. with alkaline earths and other main group metals, it is often found that the entropy term is dominant.

$$[Ca(H_2O)_6]^{2+} + edta^{4-} \rightleftharpoons [Ca(edta)]^{2-} + 6H_2O$$

| $\Delta G$/kJ mol$^{-1}$ | $\Delta H$/kJ mol$^{-1}$ | $T\Delta S$ (298K)/kJ mol$^{-1}$ |
|---|---|---|
| $-62.0$ | $-26.0$ | $36.0$ |

In some cases, the enthalpy term may actually oppose the formation of the chelated complex, although the entropy term outweighs it to give an overall favourable free energy term. In general, this situation is the exception rather than the rule.

$$[Co(H_2O)_6]^{2+} + [P_3O_{10}]^{5-} \rightleftharpoons [Co(P_3O_{10})]^{3-} + 6H_2O$$

| $\Delta G$/kJ mol$^{-1}$ | $\Delta H$/kJ mol$^{-1}$ | $T\Delta S$ (298K)/kJ mol$^{-1}$ |
|---|---|---|
| $-45.0$ | $18.8$ | $63.8$ |

## 8.2.2  Contributions to the Chelate Effect – The Enthalpy

When two ligand donor atoms are brought into proximity upon bonding to a metal ion, an electrostatic repulsion between the negative charges or dipoles is experienced. In the case of two monodentate ligands, this repulsion increases as the ligands are brought together, whereas in the case of a didentate ligand it is already 'built in' (Fig. 8-2).

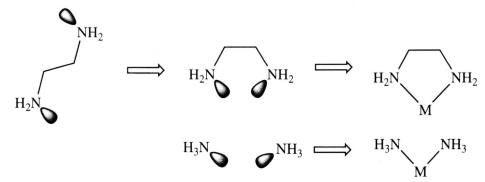

**Figure 8-2.** Schematic representation for the formation of a complex. In the case of the monodentate ligands, there is a greater unfavourable nitrogen–nitrogen repulsion involved in bringing the ligands together.

We must also consider the changes in solvation of the ligands which occur upon coordination. If we consider an amine in water, we would anticipate strong hydrogen-bonding. If we compare 1,2-diaminoethane with ammonia, we would expect the latter to be more highly solvated. This corresponds to a more unfavourable enthalpy associated with the desolvation.

### 8.2.3 Contributions to the Chelate Effect – The Entropy

The simplest way of thinking about the entropic contribution is to consider the 'half-way' stage in the formation of a chelate complex (Fig. 8-3).

**Figure 8-3.** The final ring-closure step in the formation of a chelate.

In forming the chelate complex, there is a high probability of the second donor atom Y forming a bond to the metal whereas, with monodentate ligands, the probability is much lower. In other words, once the first $M-L$ bond is formed, the second donor atom is held close to the position required for the formation of the second bond.

In more mathematical language, the favourable entropy term is associated with the release of a large number of monodentate ligands upon the formation of the chelate.

$$[Cu(NH_3)_2(H_2O)_2]^{2+} + en \rightleftharpoons [Cu(NH_3)_2(en)]^{2+} + 2H_2O$$
2 molecules                      3 molecules

| $\Delta G$/kJ mol$^{-1}$ | $\Delta H$/kJ mol$^{-1}$ | $T\Delta S$ (298K)/kJ mol$^{-1}$ |
|---|---|---|
| $-15.5$ | $-8.0$ | 8.64 |

There is also an entropy term associated with the desolvation of the ligands. This is much more difficult to assess, and may make for either favourable or unfavourable contributions to the overall entropy changes.

We now consider what ligand-field theory may contribute to an understanding of the variation in stabilities of transition-metal complexes as a function of the $d$ configuration.

## 8.3 Ligand-Field Stabilization Energies

Recall the splitting of the $d$ orbitals in octahedral environments. The energies of the $t_{2g}$ and $e_g$ subsets are shown in Fig. 8-4 with respect to their mean energy. We have used the conventional 'barycentre' formalism. In effect, we express the energy of an electron in the $t_{2g}$ or $e_g$ orbitals with respect to the total energy possessed by a set of five electrons equally distributed amongst the five $d$ functions. Alternatively, we say that our reference energy is that of a $d$ electron within the equivalent spherical mean field.

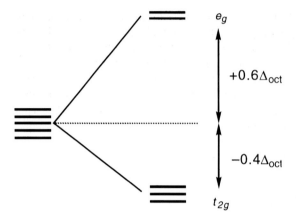

**Figure 8-4.** Splitting of the *d* orbitals in an octahedral ligand field.

In Fig. 8-5, we illustrate the orbital occupancies in the strong-field limit for $d^1$ to $d^9$ configurations in high- and low-spin arrangements. The *d*-orbital energies we associate with each $t_{2g}^n e_g^m$ configuration are computed with respect to our barycentre origin. For example, for the high-spin $d^4$ configuration in octahedral symmetry, $t_{2g}^3 e_g^1$, we add up the energies of the *d* electrons as $(-4\ Dq) + (-4\ Dq) + (-4\ Dq) + (+6\ Dq)$ to get $-6\ Dq$ (or $-0.6\ \Delta_{oct}$). We call these configuration energies *Ligand-*

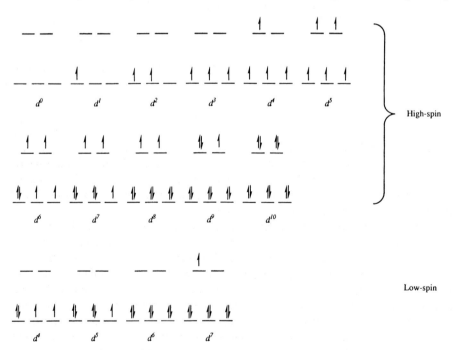

**Figure 8-5.** Electron configuration for octahedral, high- and low-spin species.

*Field Stabilization Energies*, or *LFSE*'s. The variation in LFSE across the transition-metal series is shown graphically in Fig. 8-6. It is no accident, of course, that the plots intercept the abscissa for $d^0$, $d^5$ and $d^{10}$ ions, for that is how the LFSE is defined. Ions with all other $d$ configurations are more stable than the $d^0$, $d^5$ or $d^{10}$ ions, at least so far as this one aspect is concerned. For the high-spin cases, we note a characteristic 'double-hump' trace and note that we expect particular stability conferred upon $d^3$ and $d^8$ octahedral ions. For the low-spin series, we observe a particularly stable arrangement for $d^6$ ions. More will be said about these systems in the next chapter.

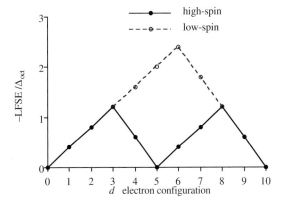

**Figure 8-6.** Comparison of LFSE terms for high- and low-spin $d^n$ configurations in octahedral ligand fields.

LFSE's for tetrahedral species are computed in a similar manner. They are compared with the results for octahedral systems in Fig. 8-7. No illustration of LFSE's for low-spin tetrahedral ions is included here because, as noted in Chapter 5, the much smaller values of $\Delta_{tet}$ relative to $\Delta_{oct}$ ensures that pairing energies $P$ always outweigh the ligand-field terms in practice.

The trends summarized in Figs. 8-6 and 8-7 arise *inevitably* from ligand-field theory. The LFSE terms are *additional* to those arising from the repulsive effects of the $d$ electrons discussed in Section 7.1. Both contributions arise simultaneously, their common origins lying in the unequal filling of the $d$-orbital subsets. To some extent, these effects interact in that the magnitudes of the LFSE terms depend directly upon the strengths of the ligand fields which themselves vary in the same qualitative manner as functions of the $t_{2g}^n e_g^m$ configurations. It is not correct, however, to argue that the variation of $Dq$ with bond length is the only consequence of the steric role of the $d$ configuration. Both $d$ electron density and bonding electron density are affected, as we have seen.

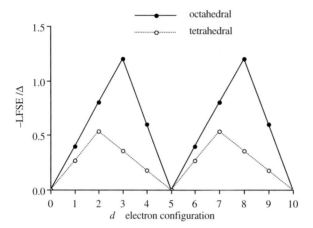

**Figure 8-7.** A comparison of high-spin octahedral and tetrahedral ligand-field stabilization energies for various $d^n$ configurations.

## 8.4 Energy and Structural Consequences in Real Systems

We shall now look at a number of thermodynamic and structural variations in transition-metal compounds which owe their origins, at least in part, to the properties of open $d$ shells.

### 8.4.1 Hydration Energies of Transition Metal(II) Ions

In Fig. 8-8 are plotted hydration energies for the first row, transition-metal divalent ions. Water is a fairly weak-field ligand and so all these $[M(H_2O)_6]^{2+}$ species are high-spin. We observe a 'double-hump' variation in $\Delta H_{hyd}$, but with respect to a gently curving, upward sloping curve. The similarlity between the relevant plot in Fig. 8-6 and the experimental data in Fig. 8-8 has long been claimed as a splendid vindication of ligand-field theory at large. And so it is – but the variations in the observed data of Fig. 8-8 compound many trends simultaneously and it is instructive to consider them all.

The enthalpies of hydration plotted in this figure refer to the process in Eq. (8.8).

$$M^{2+}(g) + aq \rightarrow [M(H_2O)_6]^{2+}(aq) \tag{8.8}$$

On the left, the divalent metal ion is spherical with a $d$-electron configuration which is amply described as $d^n$. On the right, the metal is engaged in six octahedrally disposed bonds and its $d$-electron configuration is best recorded as $t_{2g}^n e_g^m$. The electronic contributions to the hydration process refer, as usual, to the formation of the bonds and the attraction of electrons to the central metal, to the

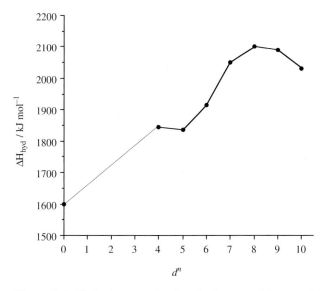

**Figure 8-8.** Hydration energies for divalent transition metal ions.

interaction of the $d$ electrons with the bonds, and to the interaction of the $d$ electrons with each other. We consider each of these in turn.

(i) Once more, let us take the extreme view that bond formation in these Wernerian complexes involves overlap of appropriate ligand orbitals with the metal $4s$ (and $4p$) but not with the $3d$ orbitals. We adopt this hueristic extreme only to emphasize the largely separate roles of the $s+p$ versus $d$ orbital sets. The metal $4s$ (and $4p$) overlap with the ligand will increase across the transition-metal period owing to the decrease in covalent or ionic radius of the metal that accompanies the increasing value of $Z_{eff}$ as discussed in Section 7.1. This effect accounts for the main underlying trend in Fig. 8-8. However, as we saw Section 7.1, the metal radii vary unevenly because of the steric effects of the $t_{2g}^n e_g^m$ configuration and so this effect already contributes to the 'double-hump' form of the experimental plot in Fig. 8-8.

(ii) Then we have the 'double-hump' contribution from the LFSE. Quantitatively, LFSE's are predicted as multiples of the ligand-field parameter $Dq$. However, $Dq$ itself varies across the period since the ligand-field strength is affected by the proximity of the bond orbitals and the $d$ orbitals. One obvious factor here is that $Dq$ should increase with decreasing bond length; another is that $Dq$ should decrease as the $3d$ orbitals contract with increasing $Z_{eff}$. Experimental $Dq$ values for divalent and trivalent first row transition-metal hexaaqua ions are plotted in Fig. 8-9. These data do seem to reflect those predicted trends somewhat. We must always remember that $Dq$ values are a complex function of many factors, however, so that the trends in Fig. 8-9 cannot be explained as simply as suggested here. Nevertheless, the dependence of the LFSE's upon $Dq$, as in Figs. 8.5–8.7, is not affected by our ability to predict the $Dq$ values themselves.

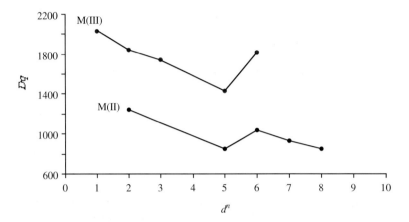

**Figure 8-9.** $Dq$ for first row transition metal $[M(H_2O)_6]^{2+}$ and $[M(H_2O)_6]^{3+}$ cations.

(iii) We must look at the contribution to the hydration energies which arises from the interactions of the $d$ electrons with each other. Behind this, in fact, is the recognition that our derivation of LFSE's took no account of such interelectron repulsion and exchange, for we implictly began the discussion at the strong-field limit. There are various ways of including the Coulomb terms and the following should be instructive.

Within the hydration process in Eqn. (8.8), a spherical ion $M^{2+}$ becomes a (hydrated) octahedral ion, $[M(H_2O)_6]^{2+}$. Part of the Coulomb energy of the free ion concerns repulsion and exchange terms within the $d^{n'}$ configuration. This is replaced by equivalent repulsion and exchange terms within the $t_{2g}^n e_g^m$ configuration. Let us estimate the trends in these quantities separately.

From the outset, we recognize that $Z_{eff}$ increases as we traverse the transition-metal period, thereby contracting the orbitals and increasing both the Coulomb and exchange integrals. Let us separate this general trend from the discontinuous ones which arise from the uneven filling of the $d$ or $t_{2g}e_g$ shells. Here, we focus upon the latter.

a) *Repulsion terms in the free ions.* Plot (a) in Fig. 8-10 indicates the variation in $d-d$ repulsion energy as the $d$ shell is progressively filled across the period. It has been constructed in recognition of the fact that all $d$ electrons repel all others; the effect is especially large when two electrons are obliged to share the same orbital.

b) *Exchange terms in free ions.* These are shown in plot (b) of Fig. 8-10. Here, the occurrence of each pair of parallel spin electrons within the same $d$ shell gives rise to an attractive exchange energy. The number of parallel-spin electron pairs in the $d^n$ configurations varies as 0, 0, 1, 3, 6, 10, 10, 11, 13, 16, 20 for n = 0 to 10 respectively.

c) *Repulsion terms in the high-spin octahedral ions.* These are estimated as above and shown as plot (c) in Fig. 8-10. At this level of approximation, they are the same as in the free ion. The fact that the $d$ orbitals are no longer equi-energetic is of no consequence here.

---

**Box 8-1**

A reminder about Coulomb and exchange integrals

Coulomb and exchange integrals arise in connection with the two-electron Coulomb operator $e^2/r_{12}$. Coulomb integrals take the form:

$$J(1,2) = \int a^*(1)b^*(2)\left(\frac{e^2}{r_{12}}\right) a(1)b(2)d\tau_1 d\tau_2 \tag{8.9}$$

$$= \int \frac{e(a^*a)_1 \cdot e(b^*b)_2}{r_{12}} d\tau_1 d\tau_2 \tag{8.10}$$

where $a$ and $b$ are the spatial parts of different orbitals. By grouping together those parts of the integrand that refer to the same variables – electrons 1 and 2 – as in Eq.(8.10), the integral $J(1,2)$ takes the form of a Coulombic repulsion between the charge clouds of orbitals $a$ and $b$ separated by distances $r_{12}$. $J(1,2)$ is called a Coulomb integral because of that.

However, because of the indistinguishability of electrons, we must also include an integral of the form:

$$K(1,2) = \int a^*(1)b^*(2)\left(\frac{e^2}{r_{12}}\right) a(2)b(1)d\tau_1 d\tau_2 \tag{8.11}$$

$$= \int \frac{e(a^*b)_1 \cdot e(b^*a)_2}{r_{12}} d\tau_1 d\tau_2 \tag{8.12}$$

On grouping functions of the same variable again, as in Eq. (8.12), we find $K(1,2)$ to describe something like the repulsion between 'overlap charge clouds'. This has no classical parallel and the integral is simply called the 'exchange' integral, for that is how it arose quantum mechanically.

Both $J$ and $K$ integrals are intrinsically positive and both vary in such a way that the closer together the electrons 1 and 2, the larger the relevant integral. As we are considering here $a$ and $b$ to be different $d$ orbitals, and as such belonging to the same atom, the magnitudes of the appropriate $J$ and $K$ integrals will be roughly comparable. Furthermore, the energies of the high-spin ground states in the species under present consideration will have the form $E = c_1 J - c_2 K$, with $c_1$ and $c_2$ positive. So, although $J$ and $K$ are intrinsically positive, while the Coulomb term gives rise to an increase in energy (repulsion), the exchange term incurs a decrease because of the minus sign (attraction). The idea that electrons can, in part, attract each other is, of course, a purely quantum phenomenon, arising ultimately from the indistinguishability of like particles.

---

d) *Exchange terms in the octahedral ions.* These are shown as plot (d) in Fig. 8-10. Now, because the $t_{2g}$ and $e_g$ orbital subsets are nondegenerate, the numbers of equivalent, parallel-spin, electron pairs are much less, as shown in Table 8-3.

Within the first-order estimations made here, it is apparent that no change in $d$–$d$ repulsion energy accompanies the hydration process. Second-order adjustments would, of course, take account of the change in mean $d$-orbital radius on complex formation. Let us agree to stop at the simple level of correction here. Overall, therefore, the significant Coulombic change on hydration concerns the loss of *exchange* stabilization.

e) *The differences in exchange terms.* Plot (e) in Fig. 8-10 represents the change in exchange energy due to the break up of degenerate parallel spins; once more, we have only made 'first-order' estimates. Here we see, yet again, a 'double-hump' contribution to the hydration energy. But we must be careful to read the plot the right way round. Plot (e) shows an increasing loss of stability following parallel-spin, electron pair disruption. So, the contribution to hydration energies will follow the trend shown in Fig. 8-11. This is the trend to be added to the $Z_{\text{eff}}$, bond length and LFSE variations. We may only guess their relative contributions. Figure 8-12 presents a possible scenario.

**Table 8-3.** The number of pairs of degenerate electrons with parallel spins in free-ion $d^n$ and octahedral $t_{2g}^n e_g^m$ configurations.

| Free ion config. | $d^0$ | $d^1$ | $d^2$ | $d^3$ | $d^4$ | $d^5$ | $d^6$ | $d^7$ | $d^8$ | $d^9$ | $d^{10}$ |
|---|---|---|---|---|---|---|---|---|---|---|---|
| Number of parallel spin pairs | 0 | 0 | 1 | 3 | 6 | 10 | 10 | 11 | 13 | 16 | 20 |
| Octahedral config. (high-spin) | $t_{2g}^0 e_g^0$ | $t_{2g}^1$ | $t_{2g}^2$ | $t_{2g}^3$ | $t_{2g}^3 e_g^1$ | $t_{2g}^3 e_g^2$ | $t_{2g}^4 e_g^2$ | $t_{2g}^5 e_g^2$ | $t_{2g}^6 e_g^2$ | $t_{2g}^6 e_g^3$ | $t_{2g}^6 e_g^4$ |
| Number of parallel spin pairs | 0 | 0 | 1 | 3 | 3 | 4 | 4 | 5 | 7 | 7 | 8 |

The plots A – F in Fig. 8-12 are defined as follows. A is the variation due to the changing $Z_{\text{eff}}$, B is that due to bond weakening from to the repulsive effects of the $t_{2g}^n e_g^m$ configuration, C is the LFSE plot of Fig. 8-6, D is a 'correction' of this for the variation in $Dq$ shown in Fig. 8-11, and E is the exchange term of Fig. 8-10. The combined contributions from A, B, D and E are plotted as F. Overall, therefore, we reproduce the form of the experimental enthalpy curve in Fig. 8-8. While satisfying, even this is not the last word. This is because we have made our estimates of the various energy contributions in terms of the strong-field limit. The hexaquo complexes lie towards the weak-field end of the appropriate correlation diagrams. The weak-field ground term for a $d^2$ octahedral complex, for example, is $^3T_{1g}(F)$. As we saw in Chapter 3, this term mixes with the excited $^3T_{1g}(P)$ term. The $^3T_{1g}(F)$ term correlates in the strong-field limit with the configuration $t_{2g}^2$ while the $^3T_{1g}(P)$ state correlates with $t_{2g}^1 e_g^1$. In an intermediate field therefore, we can describe the ground wavefunctions as combinations of $t_{2g}^2$ and $t_{2g}^1 e_g^1$ with mixing coefficients that depend upon the relative magnitudes of interelectron repulsion and the octahedral ligand field. It is certainly possible to correct our view of hydration energies in order to take all this into account: however, once again, we have no simple way of gauging the relative importance of the contributions described in Fig. 8-12. So let

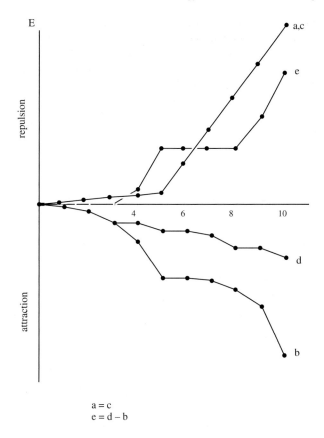

$$a = c$$
$$e = d - b$$

**Figure 8-10.** A comparison of Coulombic and exchange contributions in the free-ion and high-spin octahedral complexes (see text).

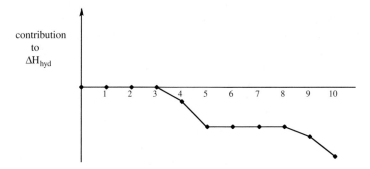

**Figure 8-11.** Contributions to $\Delta H_{hyd}$ due to disruption of numbers of degenerate, parallel electrons.

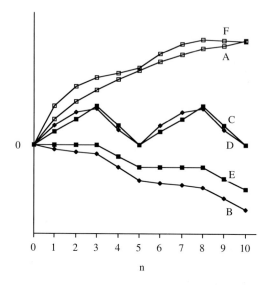

**Figure 8-12.** Contributions to heats of hydration: $A = Z_{eff}$; $B$ = bond weakening due to steric activity of $t_{2g}^n e_g^m$ configuration; $C$ = LFSE as multiples of $Dq$; $D$ = LFSE, allowing for variation in $Dq$; $E$ = exchange energy; $F = A + B + D + E$.

us be content[*] to recognize the many factors that are important while observing that the overall thermodynamic results of Fig. 8-8 are comprehensible, in principle, in terms of these factors. Let us remember also that *an explanation of the facts of Fig. 8-8 is not simply forthcoming from the LSFE plot of Fig. 8-6 alone.*

### 8.4.2  Lattice Energies of MCl$_2$ Species

In Fig. 8-13 are plotted lattice energies for MCl$_2$ species. The metal ions are high-spin and lie in octahedral sites in the lattice. The 'double-hump' form of the curve is obviously similar to that for the hydration energies we have just discussed. The reasons for the observed trend in lattice energy are virtually identical to those described for hydration energies. In one system, a metal(II) ion is octahedrally coordinated by six water molecules within a liquid medium; in the other, a metal(II) ion is octahedrally coordinated by six chlorine atoms within a solid lattice.

---

[*] Included in these would be the contributions that arise because of the different radial extents of the *d* orbitals in the free ion and complex. The factors (a) and (c) in Fig. 8-10 will not cancel exactly but, because of their general form, this 'correction' is not expected to grossly modify the *qualitative* form of plot **F** in Fig. 8-12 .

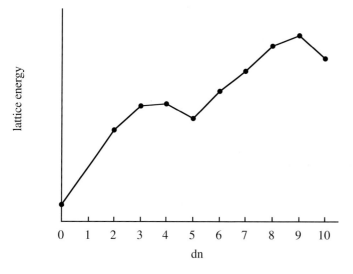

**Figure 8.13.** Lattice energies for the formation of $MCl_2$ compounds.

### 8.4.3 The Spinels

Spinels are compounds containing metals and group 16 elements of general formulae $AB_2X_4$. The $X^{2-}$ anions form an almost perfect cubic close packed array in which X may be O, S, Se or Te. By far the most numerous are the oxide spinels $AB_2O_4$. Spinel itself is $MgAl_2O_4$. The unit cell comprises 32 oxygen atoms and the formula $A_8B_{16}O_{32}$. The A metal ions are found in II, IV or VI formal oxidation states and the B metal ions then have III, II or I oxidation states respectively. Again, the most common spinels are oxide species in which A are divalent and B are trivalent metals. The cubic close packed anion lattice generates twice as many tetrahedral holes as octahedral ones. In so-called 'normal' spinels, the trivalent cations occupy half of the octahedral sites and the divalent cations one eighth of the tetrahedral sites: $A^tB_2^oO_4$. In the 'inverse' spinel structure, one half of the trivalent cations occupy tetrahedral sites, while the other half and all the divalent cations occupy octahedral sites: $B^t(AB)^oO_4$.

Typical species having the 'normal' structure are $CdAl_2O_4$, $ZnAl_2O_4$, $Mn^{II}Al^{III}_2O_4$ and $Zn^{II}Fe^{III}_2O_4$. Examples having the 'inverse' arrangement are $Ni^{II}Fe^{III}_2O_4$ and $Fe^{II}Fe^{III}_2O_4$ ($Fe_3O_4$ or magnetite). The question arises as to what factors determine the choice of 'normal' or 'inverse' structure. The most immediately obvious of these are size and charge. Generally, size considerations would predict that the smaller trivalent cations should occupy the smaller tetrahedral holes rather than the larger octahedral ones. Charge considerations, on the other hand, predict that greater lattice energies would result when the more highly charged cations are surrounded by the greater number of anions defining an octahedral hole. Straightaway, therefore, we see a conflict between these two factors and any real spinel structure involves a balance between the two. The charge factor appears to dominate in the 'normal' spinels. What has all this to do with the $d$ shell and ligand-field theory? Well, the inversion of the structure in spinels like $NiFeO_4$ is widely cited as a manifestation of the role of LFSE. The argument goes as follows.

LFSE's for high-spin octahedral and tetrahedral species were plotted in Fig. 8-9. As a direct result of the smaller values of $\Delta_{tet}$ relative to $\Delta_{oct}$, the magnitude of the stabilization energies are significantly less for the tetrahedral than for the octahedral complexes. *Other things being equal*, we deduce that LFSE's favour octahedral over tetrahedral coordination, except in $d^0$, $d^5$ or $d^{10}$ cases where both octahedral and tetrahedral LFSE's are zero. Now recall the balance of factors controlling 'normal' and 'inverse' spinel structures above and let us supplement them with the notion that the LFSE favours octahedral coordination for transition-metal ions other than those with $d^0$, $d^5$ or $d^{10}$ configurations. We observe the 'normal' structure for $Zn^{II}Fe^{III}_2O_4$ and $Mn^{II}Al^{III}_2O_4$, for example, and in these systems the transition metals $Fe^{III}$ and $Mn^{II}$ are $d^5$ species. On the other hand, in $Ni^{II}Fe^{III}_2O_4$, while $Fe^{III}$ is a $d^5$ ion, $Ni^{II}$ is $d^8$. The LSFE factor therefore favours occupancy of the octahedral sites by the nickel ions. $NiFe_2O_4$ is indeed observed to adopt the 'inverse' spinel structure, $(Fe_{0.5})^t(NiFe_{1.5})^oO_4$. A similar situation is observed in $Fe^{II}Fe^{III}_2O_4$ in which the $d^6$ iron(II) metal preferentially occupies octahedral sites. A more delicate balance occurs in $NiAl_2O_4$ in which a more nearly random site occupancy is observed: $(Al_{0.75}Ni_{0.25})^t(Ni_{0.75}Al_{1.75})^oO_4$.

The successful rationalization of these transition-metal 'inverse' spinel structures in terms of the relative LFSE's of tetrahedral and octahedral sites is another attractive vindication of ligand-field theory as applied to structure and thermodynamic properties. Once again, however, we must be very careful not to extrapolate this success. Thus, we have a clear prediction that LSFE contributions favour tetrahedral over octahedral coordination, except for $d^n$ with $n = 0$, 5 or 10. We do *not* expect to rationalize the relative paucity of tetrahedral nickel(II) species relative to octahedral ones on this basis, however. Many factors contribute to this, the most obvious and important one being the greater stabilization engendered by the formation of six bonds in octahedral species relative to only four bonds in tetrahedral ones. Compared with that, the differences in LSFE's is small beer. 'Why', one asks, 'was our rationalization of spinel structures so successful when we neglected to include consideration of the bond count?' The answer is that cancellations within the extended lattice of the spinels tend to diminish the importance of this term.

Recall that the unit cell in the spinels comprises $A_8B_{16}O_{32}$. In the 'normal' structure, there are 16 B ions in octahedral sites and 8 A ions in tetrahedral ones. That corresponds to 96 octahedral B–O bonds and 32 tetrahedral A–O bonds or 128 bonds in all. In the 'inverse' structure, we have 8 B ions in tetrahedral sites, 8 B ions in octahedral ones, and 8 A ions in octahedral sites. This corresponds to 48 octahedral B–O bonds, 32 tetrahedral B–O bonds and 48 octahedral A–O bonds or once again, 128 bonds in all. So the total number of M–O bonds, different types to be sure, is the same in both 'normal' and 'inverse' spinel structures. We could spend quite some time estimating the different bond energies of A-O and B-O or of octahedral versus tetrahedral, but that would undoubtedly involve a lot of guesswork. We can at least observe that the 'bond count' factor difference between the spinel structures is ameliorated compared with that involved with discrete octahedra and tetrahedra, and that this appears to bequeath to the LFSE a dominant role in the determination of spinel structures.


## 8.5  The Irving-Williams Series

In the late 1940's, Irving and Williams investigated the effect of varying the central metal ion on the stabilities of transition-metal complexes. Somewhat to their surprise, they found that a general pattern emerged, and that this pattern was observed both with a wide range of ligands and in the spectrum of biological activity associated with transition-metal ions. They observed that for any given ligand, the magnitude of the stability constants varied along the series:

$$K(Mn) < K(Fe) < K(Co) < K(Ni) < K(Cu) > K(Zn)$$

Some typical stability constant data are presented in Fig. 8-14, whilst Fig. 8-15 shows some biological manifestations that illustrate the ubiquity of the effect. The sequence is known as the Irving-Williams series.

Can we rationalize these observations in terms of ligand-field or other effects? The data that we have presented in Fig. 8-14 refers to the log $K_1$ values for each ligand with the high spin divalent metal ions. The sequence reflects a number of simple properties of the cations. Firstly, the trend closely parallels the 'ionic' radii

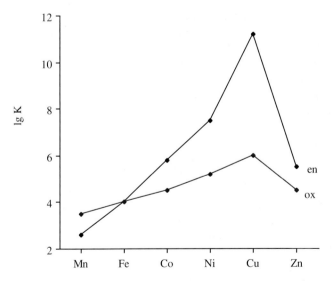

**Figure 8-14.** Variation in log $K_1$ values with metal ion for a range of divalent transition-metal ions.

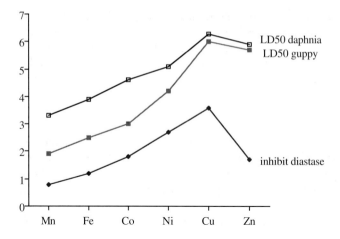

**Figure 8-15.** The biological activity of some transition-metal ions illustrating the Irving-Williams series.

of the metal ions ; as the metal ions decrease in radius, the metal–ligand interactions increase in magnitude and the stability of the complex increases. The downturn in stabilities at the end of the series is associated with the increasing ionic radii. The second correlation that we can investigate is with the LFSE associated with the electronic configuration of the metal ion. These data are also presented in Fig. 8-16, and again there is some parallel with the stability constants; the larger the LFSE, the more stable the complex.

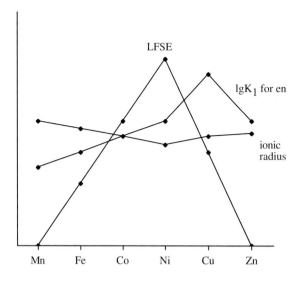

**Figure 8-16.** Correlation of ionic radius and LFSE with log $K_1$ values for divalent transition-metal complexes of 1,2-diaminoethane.

However, consideration in terms of the ionic radius or the LFSE shows that both factors predict that the maximum stabilities will be associated with nickel(II) complexes, as opposed to the observed maxima at copper(II). Can we give a satisfactory explanation for this? The data presented above involve $K_1$ values; and if we consider the case of 1,2-diaminoethane, these refer to the process in Eq. (8.13).

$$[M(H_2O)_6]^{2+} + en \rightleftharpoons [M(H_2O)_4(en)]^{2+} + 2H_2O \qquad (8.13)$$

What happens if we look at the $K_2$ or $K_3$ values for didentate ligands? In general, the $K_2$ values show stability patterns which closely parallel those for $K_1$. However, the $K_3$ values are different. Figure 8-17 presents $K_3$ data for transition-metal complexes of 1,10-phenanthroline and 1,2-diaminoethane (Eq. 8.14).

$$[M(H_2O)_6]^{2+} + 3L \rightleftharpoons [ML_3]^{2+} + 6H_2O \qquad (8.14)$$

The first feature that we note is the relative destabilization of the $[CuL_3]^{2+}$ complexes compared to the marked stabilization depicted in Fig. 8-16.

The data for the 1,2-diaminoethane complexes now parallels the trends in ionic radius and LFSE rather closely, except for the iron case, to which we return shortly. What is happening? Copper(II) ions possess a $d^9$ configuration, and you will recall that we expect such a configuration to exhibit a Jahn-Teller distortion – the six metal–ligand bonds in 'octahedral' copper(II) complexes are not all of equal strength. The typical pattern of Jahn-Teller distortions observed in copper(II) complexes involves the formation of four short and two long metal–ligand bonds.

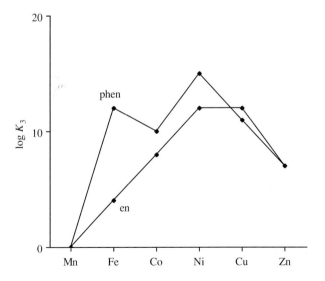

**Figure 8-17.** Stability constant data (log $K_3$) for the formation of transition-metal $[ML_3]^{2+}$ complexes.

Here is the true explanation for the position of copper(II) in the Irving-Williams series. When we consider the the replacement of water molecules by up to four other stronger-field ligands, we expect the incoming ligands to form short (and thus stronger) copper–ligand bonds. The outcome is that the Jahn-Teller distortion results in shorter and stronger metal–ligand bonds than might be expected on the basis of the isotropic 'ionic radius' of copper(II). When we come to replace the remaining two water molecules, we form metal–ligand bonds which are considerably weaker than expected. This is clearly seen when we consider the sequential data for the formation of copper(II) ammine complexes; the formation of pentammine and hexammine complexes is really very unfavourable. This is also reflected in our everyday laboratory experience – the addition of excess concentrated ammonia solution to copper(II) sulphate solutions results in the formation of the familiar deep blue solution containing the $[Cu(NH_3)_4]^{2+}$ ion rather than $[Cu(NH_3)_6]^{2+}$. This is further emphasized when we compare the sequential log $K_n$ values for copper(II) and nickel(II) ammonia complexes (Table 8-4). For the log $K_1$, log $K_2$, log $K_3$ and log $K_4$ values we see the expected Irving-Williams pattern, with the copper(II) complexes being more stable than the nickel(II) complexes. However, when we come to the log $K_5$ and log $K_6$ values, we see an inversion, with the nickel(II) complexes being considerably more stable – indeed the value of log $K_6$ for the copper(II) ammonia system cannot be measured in aqueous conditions.

**Table 8-4.** Stability constant data for copper(II) and nickel(II) ammine complexes.

|    | log$K_1$ | log$K_2$ | log$K_3$ | log$K_4$ | log$K_5$ | log$K_6$ |
|----|----------|----------|----------|----------|----------|----------|
| Cu | 4.2      | 3.5      | 2.9      | 2.1      | −0.52    |          |
| Ni | 2.8      | 2.2      | 1.7      | 1.2      | 0.7      | 0.03     |

We further emphasize this destabilization of the fifth and sixth ligands binding to copper(II) by considering the log $K_1$, log $K_2$ and log $K_3$ values for 1,2-diaminoethane complexes (Fig. 8-18). Whereas the log $K_1$ and log $K_2$ data obey the Irving-Williams sequence, the log $K_3$ parallel the trends in ionic radius and LFSE mentioned earlier. The data for *tris*(1,10-phenanthroline) complexes also illustrates the expected trend (Fig. 8-17). The anomalously high stability of the iron(II) complex can be readily explained when one considers that 1,10-phenanthroline is a very strong-field ligand and that the $[Fe(phen)_3]^{2+}$ cation is low-spin. The low-spin iron(II) centre is smaller than the high-spin analogue (0.61 Å as opposed to 0.78 Å) and has a considerably greater LFSE associated with it (24 $Dq$ as opposed to 4 $Dq$).

Our discussion of the Irving-Williams series illustrates, as ever, an important generalization in transition-metal chemistry: in many cases there is no single, simple principle which may be invoked to rationalize a given series of observations. Whilst LFSE effects are very important, they are but one of several factors controlling structure and thermodynamics.

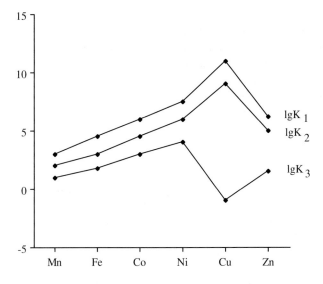

**Figure 8-18.** Stability constants for the formation of 1:1, 1:2 and 1:3 complexes with 1,2-diaminoethane.

## Suggestions for further reading

Most of the standard texts mentioned in Chapter 1 have treatments of the material discussed in this chapter. In many cases the approaches differ dramatically from that which we have adopted.

Other texts which have relevant sections include:

1. C.S.G. Phillips, R.J.P. Williams, *Inorganic Chemistry*, Oxford University Press, Oxford, **1966.**
2. K.F. Purcell and J.C. Kotz, *An Introduction to Inorganic Chemistry*, Saunders, Philadelphia, **1977.**
3. W.W. Porterfield, *Inorganic Chemistry*, Addison-Wesley, Reading (MA), **1984.**

# 9 Chemical Consequences of the *d*-Electron Configuration

## 9.1 Introduction

In the preceding chapters we have developed a detailed understanding of the behaviour of electrons in the *d* orbitals in transition-metal compounds. Can we now use this knowledge to rationalize some of the more familiar aspects of transition-metal chemistry? In this chapter we consider some of the chemical consequences of the *d*-electron configuration upon the chemistry of the transition metals. Some of the phenomena which we study are directly related to the number and arrangement of electrons in the *d* orbitals; others are *indirectly* related, being primarily dependent upon factors like the ionic radius. We address four main areas of interest to the coordination chemist: coordination number and geometry, ligand choice, oxidation state stability and rates of reactions.

## 9.2 Coordination Number and Geometry

It is reasonable to ask if it is possible to predict what the stoichiometry and geometry of the product resulting from the interaction of a particular metal ion with a particular ligand (or ligands) is likely to be. Can we make any progress towards this goal from our discussions in the earlier part of this book? As will become increasingly clear, the answer is a mixed one: sometimes the interplay of *d* electrons in the valence shell is of prime and direct importance, sometimes of little importance, but more often it is relevant, yet only in an indirect way.

The dominant features which control the stoichiometry of transition-metal complexes relate to the relative sizes of the metal ions and the ligands, rather than the niceties of electronic configuration. You will recall that the structures of simple ionic solids may be predicted with reasonable accuracy on the basis of radius-ratio rules in which the relative ionic sizes of the cations and anions in the lattice determine the structure adopted. Similar effects are important in determining coordination numbers in transition-metal compounds. In short, it is possible to pack more small ligands than large ligands about a metal ion of a given size.

This is most simply seen with monatomic ligands like the halides, which are reasonably approximated as spheres to which a meaningful radius may be assigned ($F^-$, 1.19 Å; $Cl^-$, 1.67 Å; $Br^-$, 1.82 Å; $I^-$, 2.06 Å). We should preface this discussion with the observation that, for intermediate oxidation states (+2 or +3) in donor solvents, the vast majority of complexes are based upon a six-coordinate octahedral geometry; aqueous solutions of divalent transition-metal salts all contain $[M(H_2O)_6]^{2+}$ ions. The octahedral geometry is favoured on electrostatic, ligand-field and packing grounds. We choose manganese(II), a $d^5$ ion which suffers no ligand-field imposed preference for any particular geometry, to make the point. A considerable number of complex manganese(II) fluorides, ranging from $MnF_2$ itself through $M[MnF_3]$ to $M_2[MnF_4]$, are known; all of these contain six-coordinate $MnF_6$ units. In the related chloro complexes, the octahedral structure is also the most commonly encountered, although with some M cations, four-coordinate $MnCl_4$ units are observed (remember that the radii of both the manganese and the other cation M are of importance within the crystal packing). In contrast, the bromo analogues exhibit a variety of structures in which $MnBr_6$ and $MnBr_4$ units are equally common. For example, both $K_4[MnBr_6]$ (containing a discrete $[MnBr_6]^{4-}$ anion) and $Cs_2[MnBr_4]$ (containing a discrete $[MnBr_4]^{2-}$ anion) are isolable compounds. Iodo complexes of manganese(II) are rather less common, but the majority appear to contain $MnI_4$ tetrahedra. Similar patterns exist with other $d^5$ and $d^{10}$ metal ions. For example, iron(II) forms the ion $[FeF_6]^{3-}$ with fluoride, whilst both $[FeCl_6]^{3-}$ and $[FeCl_4]^-$ are formed with chloride, and only $[FeBr_4]^-$ is known for bromide. Iron(III) iodo compounds are not commonly found since the iodide ion is a sufficiently strong reducing agent to usually reduce iron(III) to iron(II), with concomitant formation of iodine. A similar phenomenon is observed in the reaction of iodide with copper(II) salts to give copper(I) iodide and iodine. With these $d^5$ and $d^{10}$ metal ions, the influence of the electronic configuration is only indirect, through the 'ionic radius' of the metal ion.

One feature, exemplified above, is the tendency with 'borderline' ligands for metals to form stable four- and six-coordinate complexes. This is particularly marked with $d^{10}$ ions such as zinc(II), where the facile interconversion of four-, five- and six-coordinate species is believed to play an important role in the biological function of the metal in zinc metalloenzymes such as carbonic anhydrase. Further, it so happens that with $d^7$ ions such as cobalt(II) the ligand-field energies associated with a four-coordinate tetrahedral $CoL_4$ complex and an octahedral $CoL_6$ complex are similar. A common feature of cobalt(II) chemistry is the ready interconversion of these two coordination numbers. In both $d^7$ and $d^{10}$ ions, the absence of any important ligand-field 'preference' for a particular coordination number or geometry is reflected in this facile interconversion between coordination types. A familiar example of this is seen when an aqueous solution of cobalt(II) chloride is concentrated. The (dilute) pink solution contains octahedral $[Co(H_2O)_6]^{2+}$ ions. Upon concentrating, the effective concentration of chloride increases and the solution turns blue as the $[CoCl_4]^{2-}$ ion is formed. The pink colour is recovered upon dilution with water. This colour change is the basis of a very simple 'invisible' ink, a message written with dilute cobalt(II) chloride solution being invisible until the paper is warmed. The replacement of zinc(II) by cobalt(II) in zinc metalloproteins is a trick commonly used by bioinorganic chemists. The ionic radii of the two metals are somewhat similar, as is the tendency to undergo easy changes in coordination number. The success of the strategy is seen in the observation that the cobalt(II) metalloproteins very frequently show activity similar to (occasionally greater than!)

the native zinc compounds. Why should this metal-ion exchange be useful? Zinc(II) is a $d^{10}$ ion with no useful magnetic (diamagnetic) or spectroscopic (no '$d-d$' transitions) properties, whereas cobalt(II) is a $d^7$ ion with the associated paramagnetism and '$d-d$' spectra.

Size effects are probably most readily illustrated with the highly structured ligands which characterize contemporary coordination chemistry. The concept of 'cone-angle' was originally developed by Tolman to explain some of the features of phosphine coordination chemistry. The cone angle, $\theta$, was, at its simplest, defined as the angle subtended at a nickel centre (defined by a Ni–P distance of 2.28 Å) between vectors extending from the metal forming a tangent with the van der Waals extremities of the substituents on the phosphine (Fig. 9-1).

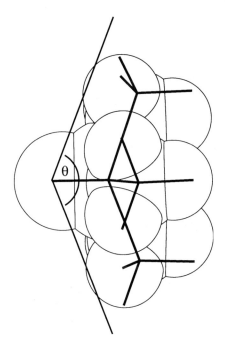

**Figure 9-1.** Definition of the cone angle, $\theta$, for trimethylphosphine.

Let us consider one specific example of how the cone-angle concept provides a good rationalization of the observed data. The reaction of nickel(II) bromide with PEtPh$_2$ gives a complex [Ni(PEtPh$_2$)$_2$Br$_2$]. This complex may be obtained as green paramagnetic or red diamagnetic forms. The two forms may be interconverted and, in solution, dynamic equilibria are set up between the two. The red form has planar geometry and the green one, tetrahedral. There is an interplay of the ligand field of the relatively strong-field P$_2$Br$_2$ donor set, which favours the formation of the square-planar complex, with the steric repulsions between the relatively bulky phosphine and halide ligands, which favour the adoption of the tetrahedral geometry in which the various ligands are kept as far apart as possible. The cone angles, $\theta$, of a number of typical phosphines are listed in Table 9-1. If the PEtPh$_2$ phosphine is replaced by PEt$_3$, with a smaller cone-angle, the percentage of the square-planar complex present at equilibrium increases, whereas replacement by PPh$_3$ increases the percentage of tetrahedral form present at equilibrium. Similarly, the complexes prepared from nickel(II) chloride favour the square-planar forms, whereas those from nickel(II) iodide favour the tetrahedral. In the solid state, the complexes with small cone-angle phosphines tend to be square-planar whereas those with the larger cone-angle ligands (generally PAr$_3$; Ar = aryl) tend to be tetrahedral.

**Table 9-1.** Cone angles for a series of phosphine ligands.

| Phosphine | $\theta$ |
| --- | --- |
| PMe$_3$ | 118 |
| PEt$_3$ | 132 |
| PEt$_2$Ph | 136 |
| PEtPh$_2$ | 140 |
| PPh$_3$ | 145 |
| P(C$_6$H$_{11}$)$_3$ | 170 |

Using *extremely* bulky ligands such as the bis(trimethylsilyl)amido anion [(Me$_3$Si)$_2$N]$^-$, it is sometimes possible to induce very low coordination numbers in transition-metal complexes. For example, a series of complexes like [M{(Me$_3$Si)$_2$N}$_3$] have been prepared. Other bulky ligands which have been used include 2,6-di-*tert*-butylpyridine, tris(mesityl)phosphine and tris(2,6-di-*tert*-butylphenyl)phosphine. The use of such bulky ligands to stabilize low coordination numbers or to stabilize highly reactive centres (by 'shielding' them from reaction) is now well developed. In general, such effects are related purely to the steric bulk of the ligands, and, once again, the only relationship to the $d^n$ configuration arises through the effective ionic radius of the metal center.[*]

A further example of ligand control of the coordination number and geometry of a complex is the use of relatively rigid polydentate ligands. For example, the phthalocyanato ligand **9.1** imposes a square-planar tetradentate N$_4$ donor set onto a metal ion, and many metals form square-planar complexes with this type of ligand. Note, however, that some metal ions also coordinate another one or two axial ligands to give square-based pyramidal or octahedral complexes.

Similarly, the macrocyclic ligand **9.2** is expected to impose a planar pentagonal N$_5$ donor set onto a metal ion. Although metal ions such as lithium form pentagonal

---

[*] A number of attempts have been made to rationalize the detailed preferences of particular $d^n$ configurations for certain geometries using molecular orbital and ligand-field based arguments. These arguments are beyond the scope of this book, and are not of general applicability to 'normal' ligands and 'normal' oxidation state metal ions.

planar complexes with this ligand, transition metals coordinate additional axial ligands to give pentagonal-based pyramidal or pentagonal bipyramidal complexes.

phthalocyaninato, **9.1**

**9.2**

It is probably true to say that we are still unable to predict the number of a given ligand which will bind to a given metal ion. However, once we know the number of ligands that bind, we may use the Kepert approach (see Chapter 1) to accurately predict the spatial arrangement of these ligands. The one exception to both of these points arises with metal ions which have a $d^8$ configuration.

We saw in Chapter 7 how the $d^8$ configuration can stabilize the square-planar arrangement of four ligands about a metal center. This is the one real success that we bring to this discussion of coordination number and geometry. In the case of first row transition-metal ions such as nickel(II), it is only very strong-field ligands which are capable of giving the necessary stabilization. We saw that weak-field ligands such as chloride give tetrahedral anions like $[NiCl_4]^{2-}$, whereas strong field species such as cyanide give square-planar ions like $[Ni(CN)_4]^{2-}$. As we descend a triad, however, the ligand-field splittings increase, with the result that nearly all palladium(II) and platinum(II) complexes are four-coordinate square-planar species. Again, note that this is *not* the case for nickel(II), where the majority of complexes possess six-coordinate octahedral geometries.

A minor success is also seen in complexes of $d^9$ and $d^4$ ions, in which the distorted octahedral geometries observed may be rationalized (and indeed predicted) in terms of the Jahn-Teller effect, and ultimately in terms of the steric activity of the open $d$ shell. This is a common feature in copper(II) chemistry, and you will recall that it was a component in our explanation of the position of copper(II) in the Irving-Williams series.

The final comment we shall make in this section concerns the formation of complexes with the low coordination numbers two and three. We remarked above that the existence of such complexes can be favoured by the use of sterically demanding ligands like the *bis*(trimethylsilyl)amido anion. Such ligands tend to be associated with transition metals to the left of the series. A second group of metal ions, but to the extreme right of the transition series, is also found to form a range of two and three coordinate complexes. This can be partially explained on electrostatic grounds. As we place more and more electrons into the $d$ manifold of the metal ion, the interelectronic repulsions between the metal ion and the ligands increase. With a full $d$ shell, these repulsions are sufficient that a range of two and three coordinate complexes like $[Ag(CN)_2]^-$ and $[CuCl_3]^{2-}$ are found with low oxidation state $d^{10}$ ions. This observation has been further rationalized in terms of the relative energies of the $ns$, $np$ and $(n-1)d$ orbitals. However, it should be noted that these low-coordinate complexes are only associated with the lower oxidation states. Zinc(II) exhibits the usual range of four- and six-coordinate complexes.

### 9.2.1  Coordination Numbers in Low Oxidation State Complexes

We have seen that complexes in low formal oxidation states (+1, zero or negative) can be stabilized by the use of strongly $\pi$-acceptor ligands like carbon monoxide or alkenes. The stabilization is associated with an increase in the ligand-field splitting resulting from the overlap of the $t_{2g}$ set of $d$ orbitals with the $\pi^*$ levels of the ligands. We also noted that these low oxidation state compounds with $\pi$-acceptor ligands are more covalent, and that the large $\Delta$ values resulted in a marked stabilization of compounds in which only the nine lowest-lying molecular orbitals are occupied – the so-called eighteen electron rule. Where the eighteen electron rule is obeyed, we can predict the number of ligands within a particular low oxidation state compound.

In many respects, the successes of this model are remarkable. Iron(0) possesses a total of eight electrons in its valence shell. To satisfy the eighteen-electron rule, five two-electron donors are needed, and compounds such as $[Fe(CO)_5]$ are formed. These molecules also obey simple VSEPR precepts, and $[Fe(CO)_5]$ adopts a trigonal bipyramidal geometry. Conversely, the use of two five-electron donor ligands such as the strong $\pi$-acceptor cyclopentadienyl, Cp, gives the well-known compound ferrocene (**9.3**).

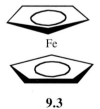

**9.3**

In a similar vein, we observe nickel(0), possessing ten electrons in its valence shell, to require four carbonyl ligands to satisfy the eighteen electron rule and form $[Ni(CO)_4]$, whilst chromium(0), with six electrons in its valence shell forms $[Cr(CO)_6]$. These latter compounds are tetrahedral and octahedral respectively.

What happens if the metal center possesses an odd number of electrons in the valence shell? Manganese(0) has seven electrons in its valence shell, and requires a total of 11 electrons to satisfy the eighteen electron rule. We can 'cheat' by using the five electron donor, Cp, and three carbonyl ligands, as in the compound $[(Cp)Mn(CO)_3]$ (**9.4**). What happens if we only have carbonyl ligands? We could form either $[Mn(CO)_5]$ (17 electrons) or $[Mn(CO)_6]$ (19 electrons), although we know that the latter species is particularly unfavourable. In fact, two things can happen. The 19 electron species $[Mn(CO)_6]$ 'wants' 18 electrons in the valence shell, and can achieve this by the loss of an electron to form the cation $[Mn(CO)_6]^+$. The alternative is for the $Mn(CO)_5$ units to dimerize, with the formation of a two-center, two-electron $Mn-Mn$ bond in $[Mn_2(CO)_{10}]$ (**9.5**) and so achieving an 18 electron configuration about each metal center.

**9.4**                                    **9.5**

When using the eighteen electron rule, we need to remember that square-planar complexes of $d^8$ centers are associated with a 16 electron configuration in the valence shell. If each ligand in a square-planar complex of a $d^8$ metal ion is a two-electron donor, the 16 electron configuration is a natural consequence. The interconversion of 16-electron and 18-electron complexes is the basis for the mode of action of many organometallic catalysts. One of the key steps is the reaction of a 16 electron complex (which is coordinatively unsaturated) with a two electron donor substrate to give an 18-electron complex.

# 9.3  Ligand Types – The Concept of Hard and Soft

One of the features of coordination chemistry which we try to explain is the 'preference' of certain metal ions for certain ligand types. In this section, we briefly discuss the models which have been developed to rationalize observed patterns of ligand recognition. We preface our discussion by noting, however, that the models

**Box 9-1**

The common ligands in low oxidation state (organometallic) chemistry and the number of electrons which they donate to the metal are indicated below. Note that the electron counting scheme that we use treats all groups as neutral. In other words, we start from neutral metal centers and treat formally anionic organic fragments as radicals. Further explanation of this point is to be found in some of the suggestions for further reading.

| Ligand | Symbol | Number of electrons |
|---|---|---|
| alkyl | R | 1 |
| aryl | Ar | 1 |
| hydride | H | 1 |
| carbonyl | CO | 2 |
| alkene | $R_2C=CR_2$ | 2 |
| phosphines | $R_3P$ | 2 |
| allyl | | 3 (or more rarely, 1) |
| diene | | 4 |
| cyclopentadienyl | | 5 (or more rarely, 3 or 1) |
| arene | ArH | 6 |

The reader is left to determine the valence shell electron count for each of the following molecules or ions: $[MeMn(CO)_5]$, $[Co_2(CO)_8]$, $[PhCr(CO)_5]^-$, $[(C_6H_6)Mo(CO)_3]$, $[(C_6H_6)_2Mo]$, $[Cr_2(CO)_{10}]^{2-}$, $[ReH_9]^{2-}$, $[(H_2CCHCH_2)Mn(CO)_5]$, $[(H_2CCHCH_2)Mn(CO)_4]$, $[(C_4H_6)Fe(CO)_3]$, $[Ti(Cp)_4]$, $[Ni(Cp)_2]$ and $[Ni(H_2CCHCH_2)_2]$

have very little to do with the *d*-electron configuration of the transition-metal ion, and more to do with the ionic size and the charge on the ion.

We begin by considering the stability constants for the formation of halide complexes with zinc(II) and mercury(II) (Table 9-2)

Notice that the stability of the zinc complexes decrease as F > Cl > Br > I, a trend that is exactly reversed for mercury. Some metals – Class (a) – form complexes

**Table 9-2.** log $K_1$ values for zinc(II) and mercury(II) halide complexes.

| M/X | F | Cl | Br | I |
|---|---|---|---|---|
| Zn(II) | 0.8 | −0.2 | −0.6 | −1.3 |
| Hg(II) | 1.0 | 6.7 | 8.9 | 12.9 |

$$M^{n+} + X^- \rightleftharpoons [MX]^{(n-1)+}$$

with stabilities decreasing in the order F > Cl > Br > I, whilst others – Class (*b*) – form complexes with stabilities increasing in the order F < Cl < Br < I.

Many metal ions parallel the behaviour of either zinc(II) or mercury(II), and Pearson described them as being Class (*a*) (hard) or Class (*b*) (soft) metals respectively. For example, iron(III) exhibits hard behaviour, whilst lead(II) is soft (Table 9-3).

**Table 9-3.** log $K_1$ values for iron(III) and lead(II) halide complexes.

| M/X | F | Cl | Br | I |
|---|---|---|---|---|
| Fe(III) | 6.0 | 1.4 | 0.5 | – |
| Pb(II) | 0.3 | 1.0 | 1.2 | 1.3 |

$$M^{n+} + X^- \rightleftharpoons [MX]^{(n-1)+}$$

Ligands which form stronger complexes with Class (a) metals are described as *hard* and those which form stronger complexes with Class (b) metals are called *soft*. *Hard metals form more stable complexes with hard ligands and soft metals form more stable complexes with soft ligands.* A listing of hard and soft metals and ligands is presented in Table 9-4.

**Table 9-4.** Hard and soft transition-metal ions and ligands.

**Hard**

Mn(II), Zn(II), Sc(III), Fe(III), Cr(III), Co(III), Ce(III), Ti(IV), Hf(IV), Zr(IV), V(IV), Mo(V), Cr(VI), W(VI), Mn(VII)
$NH_3$, $RNH_2$, $H_2O$, $HO^-$, ROH, $RO^-$, $[RCO_2]^-$, $[PO_4]^{3-}$, $[SO_4]^{2-}$, $[ClO_4]^-$, $[NO_3]^-$, $F^-$, $Cl^-$.

**Intermediate**

Fe(II), Co(II), Ni(II), Cu(II), Rh(III), Ir(III), Ru(III), Os(II)
$ArNH_2$, py, $[N_3]^-$, $Br^-$

**Soft**

M(o), Cu(I), Ag(I), Au(I), Hg(I), Cd(II), Hg(II), Pb(II), Pd(II), Pt(II)
$H^-$, $R^-$, $CN^-$, RNC, alkenes, arenes, CO, $R_3P$, $(RO)_3P$, $R_3As$, $R_3Sb$, $R_2S$, RSH, $RS^-$, $SCN^-$, $I^-$

Hard metal ions are either highly charged and/or relatively small with a high charge to radius ratio. This results in the valence shell electrons being strongly bound to the metal and less available for entering into covalent bonding with a ligand. Consequently, interaction with ligand donor atoms having high electronegativities is favoured. Hard–hard interactions are more electrostatic. In contrast, soft metal centers have low charge to radius ratios and interact with less electronegative donor atoms. Soft–soft interactions are frequently more covalent in character.

Of course, many of the ions of interest to a transition-metal chemist are 'intermediate' in character - and might do anything!

## 9.4 The Stabilization of Oxidation States, and Reduction Potentials

### 9.4.1 Reduction Potentials and Thermodynamics

In the introductory chapter we stated that the formation of chemical compounds with the metal ion in a variety of formal oxidation states is a characteristic of transition metals. We also saw in Chapter 8 how we may quantify the thermodynamic stability of a coordination compound in terms of the stability constant $K$. It is convenient to be able to assess the relative ease by which a metal is transformed from one oxidation state to another, and you will recall that the standard electrode potential, $E^{\ominus}$, is a convenient measure of this. Remember that the standard free energy change for a reaction, $\Delta G^{\ominus}$, is related both to the equilibrium constant (Eq. 9.1)

$$\Delta G^{\ominus} = -RT\ln K \tag{9.1}$$

and to the standard electrode potential (Eq. 9.2)

$$\Delta G^{\ominus} = -zFE^{\ominus} \tag{9.2}$$

where $z$ is the number of electrons involved in the redox process. From Eqs. (9.1) and (9.2), we obtain the relationship between the standard electrode potential and the stability constant for a redox process as shown in Eq. (9.3).

$$E^{\ominus} = (RT/zF)\ln K \tag{9.3}$$

We can thus use $E^{\ominus}$ values to gauge the effects that various ligands have upon the stability of one given oxidation state with respect to any other.

### 9.4.2 Intermediate Oxidation States

Negatively charged ligands are expected to stabilize higher oxidation states, and we will probe such effects shortly. Meanwhile, we may eliminate the effects of charged ligands (but not, of course, of different dipoles within a ligand) by comparing complexes with neutral ligands. Consider the cobalt complexes with six water (Eq. 9.4) and six ammonia ligands (Eq. 9.5).

$$[Co(H_2O)_6]^{3+} + e^- \rightarrow [Co(H_2O)_6]^{2+} \qquad E^{\ominus} = +1.84V \tag{9.4}$$

$$[Co(NH_3)_6]^{3+} + e^- \rightarrow [Co(NH_3)_6]^{2+} \qquad E^{\ominus} = +0.10V \tag{9.5}$$

The reduction potentials indicate[*] that the cobalt(III) aqua complex is unstable with respect to the cobalt(II) state, whereas the cobalt(III) ammine complex is

---

[*]Remember that the relevant potentials to consider are there for the oxidation and reduction of water.

stabilized. In aqueous solution, we need to consider the various redox processes by which water itself may be oxidized or reduced – for example, $[Co(H_2O)_6]^{3+}$ will oxidize water generating dioxygen and cobalt(II). Note that the large positive $E^\ominus$ for the reaction with water ligands indicates that $[Co(H_2O)_6]^{3+}$ is not likely to be an isolable species in water and, in practice, such salts may only be obtained with the greatest difficulty. Ammonia is a stronger-field ligand than water ($f$ values of 1.25 and 1.0 respectively). Cobalt(II) is a $d^7$ ion whereas cobalt(III) has a $d^6$ configuration. The somewhat stronger ligand-field of the six ammonia ligands is enough to stabilize the low spin $d^6$ configuration in the $[Co(NH_3)_6]^{3+}$ ion with its associated large LFSE. It is interesting to note that very few high-spin cobalt(III) complexes are known, and those that are possess negatively charged ligands. It is evident that in order to stabilize cobalt(III), it is necessary to have ligands which produce a sufficiently strong field to overcome the pairing energies associated with the formation of a low-spin configuration. We may usefully imagine the oxidation process to occur in two stages. Firstly, the rearrangement of the high-spin $d^7$ cobalt(II) ion to a low-spin $t_{2g}^6 e_g^1$ configuration and, secondly, the removal of an electron from the $e_g$ orbital of the cobalt(II) ion. The larger the ligand-field splitting, the greater the stabilization of the low-spin $d^6$ cobalt(III) complex. It is this ligand-field stabilization of the $d^6$ ion which compensates for the unfavourable electron repulsion associated with the low spin configuration. The subtlety of the effects involved in determining the observed $E^\ominus$ values is illustrated further by the cobalt 2,2'-bipyridine complexes (Eq. 9.6).

$$[Co(bpy)_3]^{3+} + e^- \rightarrow [Co(bpy)_3]^{2+} \qquad E^\ominus = +0.31V \qquad (9.6)$$

Here we focus upon two competing effects. The 2,2'-bipyridine is a strong-field ligand ($f = 1.33$) which will give large ligand-field splittings for both the cobalt(II) and cobalt(III) complexes. As we saw for the ammonia complexes, the splitting is sufficiently large to stabilize a low-spin cobalt(III) state. If this were the only important contribution, we would expect 2,2'-bipyridine to stabilize the cobalt(III) state more than ammonia does (since it is a stronger field ligand than ammonia). However, 2,2'-bipyridine is a $\pi$-acceptor ligand (indeed, this is the reason for its position in the spectrochemical series). The large splittings in the 2,2'-bipyridine complexes arise from the interaction of filled $t_{2g}$ orbitals on the metal with the $\pi^*$ orbitals of the ligand and the resultant lowering of the energy of the $t_{2g}$ set as shown in Chapter 6. The electron rich $d^7$ cobalt(II) ion is a better $\pi$-donor than the higher oxidation state $d^6$ cobalt(II) ion. Thus, the resultant lowering of the $t_{2g}$ orbitals will be more effective with the cobalt(II) than the cobalt(III) ion. The balance is such that the stabilization of the cobalt(III) state is less with 2,2'-bipyridine than with the weaker-field $NH_3$ ligand! The paradox is that we would normally expect larger ligand-field effects to be associated with the higher oxidation state. We investigate this in a little more detail by studying some iron(II)/(III) complexes in which the $d^6$ configuration is associated with the *lower* oxidation state.

$$[Fe(H_2O)_6]^{3+} + e^- \rightarrow [Fe(H_2O)_6]^{2+} \qquad E^\ominus = +0.77V \qquad (9.7)$$

$$[Fe(bpy)_3]^{3+} + e^- \rightarrow [Fe(bpy)_3]^{2+} \qquad E^\ominus = +0.97V \qquad (9.8)$$

Consider, then, the iron complexes with water (Eq. 9.7) and 2,2'-bipyridine ligands (Eq. 9.8). The larger $E^{\ominus}$ with the 2,2'-bipyridine ligand indicates a greater stabilization of the iron(II) complex with the stronger-field $\pi$-acceptor ligand. This is for exactly the same reasons we discussed when comparing the cobalt complexes. In this case, the *lower* oxidation state is stabilized by the 2,2'-bipyridine ligand. This is because of the favourable ligand-field terms associated with the $d^6$ configuration with the strong field ligand. In the iron complexes, this stabilizes the iron(II) state whereas in the cobalt complexes, it is the cobalt(III) state which benefits and in fact, the blue $[Fe(bpy)_3]^{3+}$ ion is not particularly stable in water. Even this is not the whole story for we are not really comparing like with like – the 2,2'-bipyridine complexes of iron are low-spin in both the iron(II) and iron(III) states whereas the aqua complexes are high spin. So we see the origin of an additional stabilization of the $d^6$ $[Fe(bpy)_3]^{2+}$ complex ion.

As expected, the introduction of negatively charged ligands results in the stabilization of the higher oxidation states. This is seen most simply in the comparison of aqua and oxalato complexes of cobalt (Eqs. 9.9 and 9.10). Oxalate is comparable in ligand-field strength to water ($f = 0.99$) but the negatively charged ligands stabilize the higher oxidation state. All of our remarks regarding the change from the high-spin cobalt(II) to the low-spin cobalt(III) ion pertain here.

$$[Co(H_2O)_6]^{3+} + e^- \rightarrow [Co(H_2O)_6]^{2+} \qquad E^{\ominus} = +1.84V \qquad (9.9)$$

$$[Co(ox)_3]^{3-} + e^- \rightarrow [Co(ox)_3]^{4-} \qquad E^{\ominus} = +0.57V \qquad (9.10)$$

$$[Co(edta)]^- + e^- \rightarrow [Co(edta)]^{2-} \qquad E^{\ominus} = +0.6V \qquad (9.11)$$

A similar stabilization of the cobalt(III) state is observed if we use a chelating ligand such as edta$^{4-}$ (Eq. 9.11), which completely 'wraps-up' the metal center. Here is a most important point. Why do negatively charged ligands stabilize higher oxidation states? It has more to do with the entropy term associated with the solvation of the more highly charged ions than the enthalpy term reflecting any differences in M–L bond strengths.

What happens if we 'boost' the effect of negatively charged ligands by choosing one which is also a strong-field $\pi$-acceptor? A good example is provided by the cyano complexes (Eq. 9.12). Note that in this case, the high-spin square-based pyramidal cobalt(II) ion is only coordinated to five cyanide ligands.

$$[Co(CN)_6]^{3-} + e^- \rightarrow [Co(CN)_5]^{3-} + CN^- \qquad E^{\ominus} = -0.83V \qquad (9.12)$$

This may be rationalized in terms of two factors. Firstly, the tendency to build up a large charge density on the cobalt(II) center would be great (but remember the electroneutrality principle) and, secondly, the ligands are labilized by the presence of electron density in the $e_g$ orbitals of the high-spin cobalt(II) ion. The loss of one negatively charged ligand is not possible in the chelated oxalato or edta complexes (Eqs. 9.10 and 9.11). The massive stabilization of the cobalt(III) state is partially due to the negatively charged ligands and partly due to the ligand-field stabilization of the low-spin $d^6$ ion.

Similar effects are observed in the iron complexes of Eqs. (9.13) and (9.14). The charge on the negatively charged ligands dominates the redox potential, and we observe stabilization of the iron(III) state. The complexes are high-spin in both the oxidation states. The importance of the low-spin $d^6$ configuration (as in our discussion of the cobalt complexes) is seen with the complex ions $[Fe(CN)_6]^{4-}$ and $[Fe(CN)_6]^{3-}$ (Eq. 9.15), both of which are low-spin.

$$[Fe(ox)_3]^{3-} + e^- \rightarrow [Fe(ox)_3]^{4-} \qquad E^\ominus = +0.02V \qquad (9.13)$$

$$[Fe(edta)]^- + e^- \rightarrow [Fe(edta)]^{2-} \qquad E^\ominus = -0.12V \qquad (9.14)$$

$$[Fe(CN)_6]^{3-} + e^- \rightarrow [Fe(CN)_6]^{4-} \qquad E^\ominus = +0.36V \qquad (9.15)$$

The now-familiar balance of effects operates. The anionic ligands favour the higher oxidation state (again, associated with a solvation effect), the ligand-field stabilization of the low-spin $d^6$ iron(II) center is considerably greater than that of the low-spin $d^5$ iron(III) center, and back-donation is expected to be greater in the iron(II) complex. The latter is indeed the case for the Fe–C distances in the iron(II) compound are slightly shorter than those in the iron(III). The overall balance is a stabilizing of the iron(III) state with respect to complexes with aqua ligands, in contrast to those with the neutral strong-field ligand bpy.

### 9.4.3 The Electroneutrality Principle - A Reprise

In Chapter 1 we introduced the electroneutrality principle. We now consider some of its implications. You will recall that we described the $[Fe(H_2O)_6]^{3+}$ ion as 50% covalent (or ionic). A similar description of 50% covalency may be applied to the cobalt(III) complex ions $[Co(H_2O)_6]^{3+}$ and $[Co(NH_3)_6]^{3+}$. The higher the oxidation state, the greater the covalency necessary in the bond to fulfill the requirements of the electroneutrality principle. The electronegativities of cobalt(III), N and O are 2.0, 3.0 and 3.4 respectively. The smaller difference in electronegativities between cobalt(III) and nitrogen than between cobalt(III) and oxygen means that the Co–N bond will be more covalent than the Co–O bond. In accord with the requirements of the electroneutrality principle, $[Co(NH_3)_6]^{3+}$ will be more favoured than $[Co(H_2O)_6]^{3+}$. This is one of the observations that we discussed in the previous section. Note that we have come to the same conclusion without invoking any knowledge of the number or arrangement of the $d$ electrons.

Let us extend our discussion of the $[Fe(H_2O)_6]^{3+}$ cation a little further. The electronegativities of oxygen and hydrogen are 3.4 and 2.2 respectively, and the O–H bond should thus be polarized in the sense $H^{\delta+}$–$O^{\delta-}$. The electroneutrality principle applied to the Fe–O interaction resulted in our placing half-positive charges upon each of the oxygen atoms. If we now consider the H–O interactions, we may reallocate charges in accord with the electronegativities and the electroneutrality principle such that the positive charges reside on the hydrogens. Charge neutrality of the oxygen would then establish a final charge distribution which places a quarter positive charge on each hydrogen atom, and each oxygen atom and the iron center are neutral. The overall charge on the complex ion ($12 \times 1/4 = 3$) is thus, of course, unchanged.

Consider the closely related ion $[Fe(H_2O)_6]^{2+}$. The only difference is in the formal oxidation state of the metal ion. If an ionic model is assumed (**9.6**), the charge on the metal center is +2. A purely covalent model results in the placing of a formal quadruple negative charge upon the iron center (**9.7**). To satisfy the electroneutrality principle, and establish a near-zero charge on the metal, each oxygen atom is

**9.6**

**9.7**

required to donate 1/3 of an electron (**9.8**). Pauling describes this situation as being 33% covalent (or 66% ionic). We could envisage further distribution of the electronic charge such that both the iron and the oxygen atoms are neutral, so giving a 1/6 positive charge associated with each hydrogen atom. We shall return to this observation shortly. Remember, meanwhile, that there is a smaller positive charge assigned to the hydrogen atoms in $[Fe(H_2O)_6]^{2+}$ than in $[Fe(H_2O)_6]^{3+}$.

**9.8**

Now consider an iron(III) complex with six negatively charged ligands. The purely covalent representation places a $-3$ charge on the metal center and leaves each ligand neutral. The electronegativity principle makes the iron neutral and places a half-negative charge upon each ligand. Ligand donor atoms are invariably more electronegative than metal centers, and so this distribution of charges is favoured. Compare this situation with that in the $[Fe(H_2O)_6]^{3+}$ ion in which we ended up placing positive charge upon the ligands. This phenomenon, which ultimately rests upon the typical relative electronegativities of the metal and ligands, provides a second main cause of the stabilization of higher oxidation states by negatively charged ligands.

## 9.4.4 Protic Equilibria Involving Coordinated Ligands

We have just noted that the electroneutrality principle suggests that, in cationic aqua complexes, the hydrogen atoms of the water ligands acquire positive charge. Furthermore, we observed that the higher the oxidation state, the greater the positive charge and the greater the polarization of the O–H bonds. In other words, the higher the oxidation state, the more acidic the water ligands become and equilibria of the type shown in Eq. (9.16) become accessible.

$$[L_5M(OH_2)]^{n+} \rightleftharpoons [L_5M(OH)]^{(n-1)+} + H^+ \qquad (9.16)$$

This is indeed observed and, particularly in higher oxidation states, coordinated water molecules are relatively acidic (Table 9-5). Water coordinated to an iron(III) center is a stronger acid than acetic acid!

**Table 9-5.** $pK_a$ values for coordinated water molecules.

|  | $pK_a$ |
| --- | --- |
| $H_2O$ | 15.6 |
| $[Al(H_2O)_6]^{3+}$ | 5.0 |
| $[Fe(H_2O)_6]^{3+}$ | 2.0 |
| $[Zn(H_2O)_6]^{2+}$ | 9.5 |

If we combine this observation with the previous discussion regarding the use of negatively charged ligands to stabilize higher oxidation states, we have a self-regulating way in which aqua ions may 'adjust' their coordination environment as the oxidation state of the central metal ion changes. The higher the oxidation state of the metal ion, the greater the polarization of the water molecule and the more acidic it becomes; the more acidic the water, the greater the tendency to form hydroxide (or even oxide) ligands which then stabilize the high oxidation state of the metal ion.

Consider some vanadium ions in aqueous solution. Pale violet solutions of vanadium(II) salts contain the $[V(H_2O)_6]^{2+}$ ion. The vanadium(II) center is only weakly polarizing, and the hexaaqua ion is the dominant solution species. Aqueous vanadium(II) solutions are observed to be unstable with respect to *reduction* of water by the metal center. In contrast, vanadium(III) is more highly polarizing and an equilibrium between the hexaaqua and pentaaquahydroxy ion is set up. The $pK_a$ of 2.9 means that the $[V(OH_2)_6]^{3+}$ ion (Eq. 9.17) only exists in strongly acidic solution or in stabilizing crystal lattices.

$$[(H_2O)_5V(OH_2)]^{3+} \rightleftharpoons [(H_2O)_5V(OH)]^{2+} + H^+ \qquad pK_a = 2.9 \qquad (9.17)$$

Vanadium(IV) is even more strongly polarizing. The first deprotonation process is not observable in aqueous solution. The pentaaquahydroxy ion *may* be present in very strongly acidic solutions, but the dominant solution species is one which is derived from a second deprotonation of the pentaaquahydroxy complex. This could give rise to a bis(hydroxy)tetraaqua ion or an oxopentaaqua species. We will see later that the multiply bonding $\pi$-donor oxo group stabilizes higher oxidation state ions, and it is, in fact, this latter species which is present (Eq. 9.18).

$$[(H_2O)_5V(OH_2)]^{4+} \rightleftharpoons [(H_2O)_5V(OH)]^{3+} + H^+$$

(9.18)

$$[(H_2O)_5V(OH)]^{3+} \rightleftharpoons [(H_2O)_5V(=O)]^{2+} + H^+$$

The blue $[(H_2O)_5V(=O)]^{2+}$ ion is the vanadyl ion which is usually depicted as $VO^{2+}$. Actually, the vanadium center is still sufficiently polarizing that a third deprotonation equilibrium is established in aqueous media to generate the ion $[(H_2O)_4V(=O)(OH)]^+$, which contains water, hydroxy and oxo ligands (Eq. 9.19).

$$[(H_2O)_4V(=O)(H_2O)]^{2+} \rightleftharpoons [(H_2O)_4V(=O)(OH)]^+ + H^+ \qquad pK_a \; 6.0 \qquad (9.19)$$

A similar situation pertains for iron salts in aqueous solution. Solutions of iron(II) salts contain the very pale green cation $[Fe(H_2O)_6]^{2+}$, although these solutions often appear with various darker shades as a result of aerial oxidation. In the solid state, the alum $KFe(SO_4)_2 \cdot 12H_2O$ is a very pale violet colour, and contains the $[Fe(H_2O)_6]^{3+}$ ion. Solutions of this compound or other iron(III) salts are usually varying shades of yellow, although very pale coloured solutions may be obtained in acidic conditions. The yellow coloration is due to the deprotonated species which exhibit a ligand–metal-charge transfer transition in the ultraviolet region which tails into the visible. Both mono- (Eq. 9.20) and bis-deprotonated (Eq. 9.21) complexes are present in aqueous solution. Note the difference between the iron(III) and the higher oxidation state vanadium(IV) complexes. In the latter case, an oxo ligand was generated after the second deprotonation to stabilize the high oxidation state metal centre, whereas with the lower oxidation state iron(III) centre, a bis(hydroxy) complex is formed (Eq. 9.21).

$$[(H_2O)_5Fe(OH_2)]^{3+} \rightleftharpoons [(H_2O)_5Fe(OH)]^{2+} + H^+ \qquad pK_a = 2.0 \qquad (9.20)$$

$$[(H_2O)_4Fe(OH_2)(OH)]^{2+} \rightleftharpoons [(H_2O)_4Fe(OH)_2]^+ + H^+ \qquad pK_a = 3.3 \qquad (9.21)$$

This leads us to a second aspect of the formation of hydroxy ligands in higher oxidation state complexes. The deprotonation of a coordinated water ligand to a hydroxy ligand is frequently associated with the formation of polynuclear complexes in which the hydroxy ligands are associated with two metal ions which they bridge, as opposed to a single metal center. This is a process which is known as *olation* and was first described by Werner in his pioneering studies of kinetically inert cobalt(III) complexes.

The vanadium(III) ion $[V(H_2O)_6]^{3+}$ (**9.9**) exhibits this behaviour, with a log $K$ of 4 associated with the formation of the hydroxy-bridged dinuclear complex **9.10**. This is a general phenomenon. For example, chromium(III) and iron(III) form strictly

**9.9**                                      **9.10**

analogous olated dinuclear complexes. In the case of the iron(III) species, the process can proceed further to generate bridging oxy ligands (Eqs. 9.22–9.24).

$$2[(H_2O)_5Cr(OH_2)]^{3+} \rightleftharpoons [(H_2O)_4Cr(OH)_2Cr(H_2O)_4]^{4+} + 2H^+ \qquad \log K = 4.0 \qquad (9.22)$$

$$2[(H_2O)_5Fe(OH_2)]^{3+} \rightleftharpoons [(H_2O)_4Fe(OH)_2Fe(H_2O)_4]^{4+} + 2H^+ \qquad \log K = 3.3 \qquad (9.23)$$

$$[(H_2O)_4Fe(OH)_2Fe(H_2O)_4]^{4+} \rightleftharpoons [(H_2O)_4Fe(O)_2Fe(H_2O)_4]^{2+} + 2H^+ \qquad (9.24)$$

This is only the beginning of a process which ultimately results in the formation of solid state hydroxides or oxides. Actually, the solution species present in neutral or alkaline solutions of transition-metal ions are relatively poorly characterized. The formation of numerous hydroxy- and oxy-bridged polynuclear species makes their investigation very difficult. However, it is clear that there is a near-continuous transition from mononuclear solution species, through polynuclear solution species to colloidal and solid state materials. By the way, the first example of a 'purely' inorganic compound to exhibit chirality was the olated species **9.11**.

**9.11**

## 9.4.5  The Stabilization of High Oxidation States

There is an interesting paradox in transition-metal chemistry which we have mentioned earlier – namely, that low and high oxidation state complexes both tend towards a covalency in the metal–ligand bonding. Low oxidation state complexes are stabilized by $\pi$-acceptor ligands which remove electron density from the electron rich metal center. High oxidation state complexes are stabilized by $\pi$-donor ligands which donate additional electron density towards the electron deficient metal centre.

The stabilization of high oxidation state compounds might seem particularly paradoxical. The highest oxidation states are usually only stabilized by ligands such as fluoride and oxide. Complex species such as $[MnO_3F]$ and $[Mn_2O_7]$ represent the stabilization of manganese(VII), $[CrO_4]^{2-}$, $[Cr_2O_7]^{2-}$ and $[CrO_3Cl]^-$ of chromium(VI), whilst iron(VI) is observed in $Na_2[FeO_4]$. The oxo ligands form formal double bonds with the metal, and the short $M-O$ distances that result allow efficient transfer of charge to the electron deficient metal centre. In the case of fluoride, the short $M-F$ distances (as fluoride is a relatively small ligand) allow efficient overlap between the filled $2p$ orbitals of the fluorine and the empty orbitals of the metal. And yet a further paradox: the ligands which stabilize the highest oxidation states are those with the most electronegative donor atoms! However, despite the fluorine being electronegative, it is acting as a $\pi$-donor to the metal!

## 9.4.6  The *d* Orbitals, Covalent Character and Variable Oxidation States – A Summary

In Chapter 6, we introduced the idea of the variable role of the *d* orbitals in transition-metal complexes as a function of changing oxidation state. At that point, we focused upon the difference between low oxidation states and 'higher' ones (meaning those typical of Werner-type compounds; say, +2 or +3). In *this* chapter, we have concentrated rather more on high oxidation states and have noted the 'paradoxical' variations in the covalent character of the $M-L$ bonds with varying oxidation state. We now draw these various themes together and provide an overview of changing bonding character throughout the *d*-block chemistry.

In Fig. 9-2, we offer a schematic summary of the determinants of covalent character in transition-metal bonding.

In intermediate and high oxidation states, the (3)*d* orbitals are increasingly 'inner' with respect to the (4)*s* and (4)*p* orbitals. As discussed repeatedly throughout this book, these *d* orbitals may be considered as largely uninvolved in direct overlap with the ligand orbitals: they are essentially excluded from the metal's valence shell. In changing from, say, the +2 oxidation state to the +7 state, the polarizing power, or hardness, of the notional $M^{n+}$ ion increases dramatically and non-linearly (curve 1 in Fig. 9-2). Higher oxidation states will be accessible only with hard ligands (soft anions would reduce the metal in these higher oxidation states). Though retained tightly by the ligands, their electron density is drawn towards the more highly charged metal and so the covalent character of the $M-L$ bonds increases steadily with increasing $n$. The same conclusion follows with the recognition that

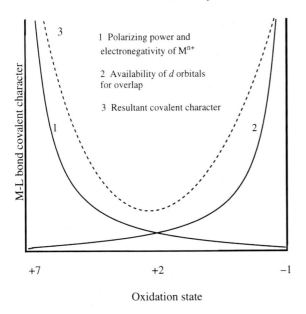

**Figure 9-2.** The variation of covalent character with oxidation state.

the electronegativity of a progressively more positive metal ion rapidly approaches, or even surpasses, that of even an atom like fluorine. So the (perhaps) 50% covalent character of the M−L bonds in typical metal(II) complexes increases with increasing oxidation state.

On the other hand, a decrease in oxidation state from metal(II) quite rapidly releases the $(3)d$ orbitals into the valence shell. The $d$ orbitals are full, or frequently so, so that M−L bonding electron density now derives from both metal and ligand (this is in contrast to the dative covalency of higher oxidation state complexes in which the electron density emanates from the ligands alone). We thus observe an increase in covalent character on decreasing the oxidation state from metal(II), but this time originating from the rapidly increasing participation of the $(3)d$ orbitals in the valence shell (and the emerging dominance of the 18-electron rule). Towards this extreme, we see the growing importance of soft metal/soft ligand interactions, mediated by the synergic 'back-bonding', first mooted by Chatt, Dewar and Duncanson.

Throughout the scheme summarized above, we are to understand that multiple bonding involves $(3)d_\pi-L_\pi$ overlap in the low oxidation state complexes, but $(4)p_\pi-L_\pi$ overlap in the high oxidation state complexes. These latter are generally characterized by substantially shorter bonds, thus facilitating $p_\pi-L_\pi$ overlap relative to a presumably small such contribution in less tightly bound Werner-type systems. In any case, the degree of $(4)p_\pi-L_\pi$ overlap need not be great since all that is required to satisfy the electroneutrality principle is a sufficient drift of electron density towards the metal. This could be dominated by $\sigma$-bonding contributions although we need not guess the relative proportions of $\sigma$ and $\pi$, however, to make the main point.

## 9.5  Consequences of the *d*-Electron Configuration upon Reaction Rates

Thus far, we have focused upon the thermodynamic consequences of the *d*-electron configuration. Many everyday observations in transition-metal chemistry have more to do with the relative *rates* of reactions rather than the position of a thermodynamic *equilibrium*. So now we consider some of the kinetic manifestations of partially filled *d* orbitals.

### 9.5.1  Kinetically Inert and Labile Complexes

It is convenient to divide the discussion of the mechanistic behaviour of transition-metal complexes into those of labile and non-labile complexes, imperfect though this division may be. The description of a complex as labile or non-labile is empirical, being based upon the typical time it takes for a reaction to proceed to completion. We adopt the suggestion of Taube, which refers to substitution reactions in which one of the ligands coordinated to a metal center is replaced by another ligand. If this process is complete in less than one minute (at 298 K with reactant concentrations of 0.1 M), then the complex is described as labile, whereas if it takes considerably longer than this time, the complex is described as non-labile or inert. In what follows, we should note two generalities. Firstly, inert complexes are not necessarily thermodynamically stable with respect to the reaction under consideration; conversely, thermodynamically stable complexes often undergo rapid reactions. Secondly, the properties of lability and inertness are found to be loosely associated with particular metal ions in particular oxidation states; complexes of cobalt(III), chromium(III) and most second and third row transition metals are generally inert.

### 9.5.2  Ligand Substitution Reactions

One of the commonest reactions in the chemistry of transition-metal complexes is the replacement of one ligand by another ligand (Fig. 9-3) – a so-called substitution reaction. These reactions proceed at a variety of rates, the half-lives of which may vary from several days for complexes of rhodium(III) or cobalt(III) to about a microsecond with complexes of titanium(III).

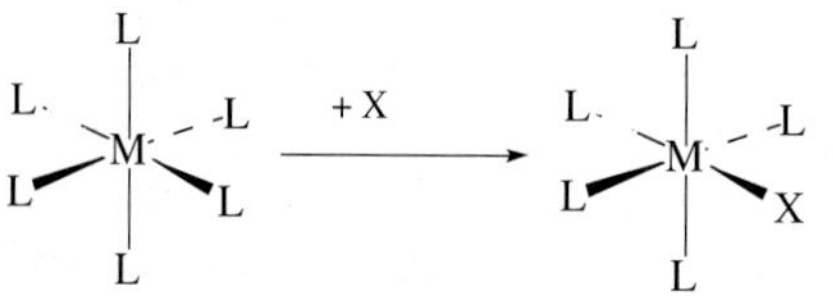

**Figure 9-3.** The substitution of L by X in an octahedral complex.

The precise mechanism by which this process occurs has been the subject of considerable study and debate over the past thirty years. Limiting mechanisms would involve an associative process in which an intermediate or transition state of increased coordination number is formed (the $S_N2$ mechanism of organic chemistry represents a limiting associative process), or a dissociative one with an intermediate or transition state of lower coordination number (the limiting $S_N1$ mechanism of organic chemistry). These mechanisms differ in the relative importance of bond-making to the incoming ligand and bond-breaking with the leaving ligand in the transition state. In general, the mechanisms are thought to be of the *interchange* type in which bond-making with the incoming group is concurrent with bond-breaking to the leaving group. These reactions are delineated $I_a$ or $I_d$ depending whether bond-making to the incoming ligand or bond-breaking to the leaving ligand is considered to be dominant in the transition state. In many cases, conventional kinetic studies do not provide data to allow unambiguous assignment of the mechanism for substitution reactions. The reader is referred to the reading list at the end of this chapter for further information upon this vexing subject!

As already mentioned, complexes of chromium(III), cobalt(III), rhodium(III) and iridium(III) are particularly inert, with substitution reactions often taking many hours or days under relatively forcing conditions. The majority of kinetic studies on the reactions of transition-metal complexes have been performed on complexes of these metal ions. This is for two reasons. Firstly, the rates of reactions are comparable to those in organic chemistry, and the techniques which have been developed for the investigation of such reactions are readily available and appropriate. The time scales of minutes to days are compatible with relatively slow spectroscopic techniques. The second reason is associated with the kinetic inertness of the products. If the products are non-labile, valuable stereochemical information about the course of the substitution reaction may be obtained. Much is known about the stereochemistry of ligand substitution reactions of cobalt(III) complexes, from which certain inferences about the nature of the intermediates or transition states involved may be drawn. This is also the case for substitution reactions of square-planar $d^8$ complexes of platinum(II), where study has led to the development of rules to predict the stereochemical course of reactions at this centre.

It will not have escaped the reader's attention that the kinetically inert complexes are those of $d^3$ (chromium(III)) or low-spin $d^6$ (cobalt(III), rhodium(III) or iridium(III)). Attempts to rationalize this have been made in terms of ligand-field effects, as we now discuss. Note, however, that remarkably little is known about the nature of the transition state for most substitution reactions. Fortunately, the outcome of the approach we summarize is unchanged whether the mechanism is associative or dissociative.

Basolo and Pearson, in their classical work on inorganic reaction mechanisms, developed a ligand-field based approach to understanding the occurrence of kinetically inert transition-metal ions. They calculated the LFSE associated with the starting (octahedral) complex for a given $d^n$ complex ($Dq$ values were deduced from actual spectroscopic data). They then considered limiting associative and disociative mechanisms leading to seven- or five-coordinate intermediates. The five-coordinate intermediate in the dissociative process might exist as a square-based pyramid or as a trigonal bipyramid. They attempted then to determine the ligand-field stabilization energy for the intermediate by assuming that the overall magnitude of the ligand-field from the ligands is unchanged on passing from the starting complex to the intermediate. Although the justification for this may seem tenuous, it being argued that the change in number of ligands is countered by changes in bond lengths, recent and detailed ligand-field analyses support this early idea. For example, the reduction in the ligand-field expected on going from six to five ligands is balanced by the five ligands being closer to the metal ion in response to the requirements of the electroneutrality principle. The ligand-field splittings for a variety of geometries were so calculated (in terms of $\Delta_{oct}$). The change in LFSE between starting complex and intermediate was termed the ligand-field activation energy (LFAE). A decrease in ligand-field stabilization energy upon passing from the ground state to the transition state (a positive LFAE in their definition), would provide an additional contribution to the overall activation energy for the substitution process. A negative contribution corresponds to a lowering of the activation energy for substitution. Basolo and Pearson found that the LFAE for substitution of $d^3$ or low-spin $d^6$ was positive regardless of the coordination number or geometry of the transition state. [Perhaps this is just another way of stating that the LFSE for octahedral $d^3$ or low-spin $d^6$ centers is high!] Clearly, the approach hinges upon the reliability of the estimates of the ligand-field stabilization energy for the transition state, whose detailed geometry is unknown.

---

**Box 9-2**

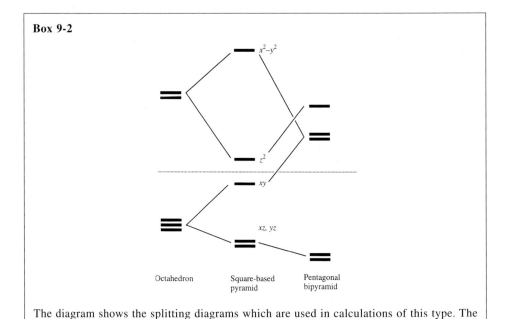

Octahedron          Square-based          Pentagonal
                    pyramid               bipyramid

The diagram shows the splitting diagrams which are used in calculations of this type. The LFSE for an octahedral $d^3$ ion is $-1.2\Delta_{oct}$. The estimated LFSE's for square-planar five coordinate and pentagonal bipyramidal seven-coordinate transition states are $-1.0\Delta_{oct}$ and $-0.774\Delta_{oct}$ respectively, leading to LFAE's of $+0.2\Delta_{oct}$ and $+0.426\Delta_{oct}$ respectively.

### 9.5.3  Rates of Electron Transfer Reactions

In the same way that we considered two limiting extremes for ligand substitution reactions, so may we distinguish two types of reaction pathway for electron transfer (or redox) reactions, as first put forth by Taube. For redox reactions, the distinction between the two mechanisms is more clearly defined, there being no continuum of reactions which follow pathways intermediate between the extremes. In one pathway, there is no covalently linked intermediate and the electron just "hops" from one center to the next. This is described as the *outer-sphere* mechanism (Fig. 9-4).

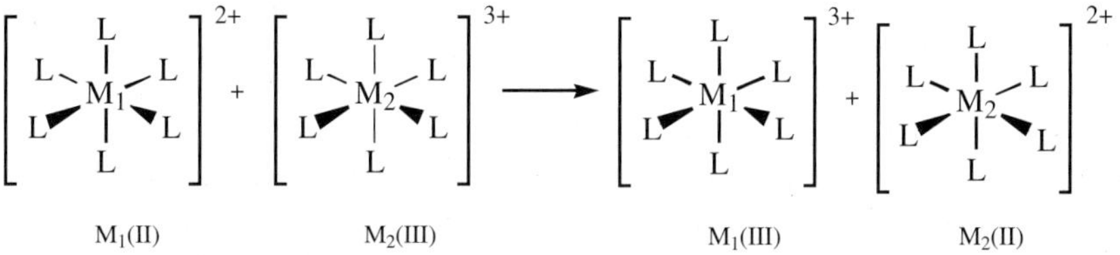

**Figure 9-4.** The outer-sphere mechanism for an electron transfer reaction between two complexes. No covalently-linked intermediate is involved in the reaction.

The second mechanism involves the formation of a covalent bridge through which the electron is passed in the electron transfer process. This is known as the *inner-sphere* mechanism (Fig. 9-5).

The inner-sphere mechanism is restricted to those complexes containing at least one ligand which can bridge between two metal centers. The commonest examples of such ligands are the halides, hydroxy or oxo groups, amido groups, thiocyanate

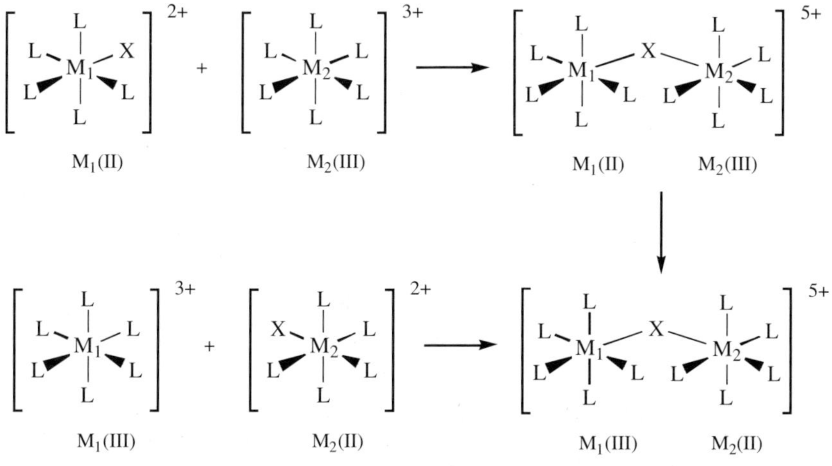

**Figure 9-5.** The inner-sphere mechanism for an electron transfer reaction between two complexes. A covalently-linked intermediate *is* involved in this reaction.

and more complex conjugated organic ligands such as pyrazine (**9.12**) or 4,4'-bipyridine (**9.13**).

**9.12**                                    **9.13**

The scheme in Fig. 9-5 above illustrates the case in which the bridging ligand, X, is transferred from metal center $M_1$ to $M_2$ in the course of the reaction. Although this is not a necessary consequence of an inner-sphere pathway, it is often observed, and provides one method for establishing the mechanism.

It is often very difficult to distinguish one mechanism from another, but some clever experiments based upon LFSE effects have been designed. In the previous section, we established that $d^3$ and low-spin $d^6$ metal complexes are kinetically inert, and only undergo ligand substitution and displacement reactions with difficulty. Study of electron transfer reactions between two such kinetically inert metal centers suggests that the redox processes proceed by outer-sphere mechanisms, since we cannot form the new metal-ligand bonds necessary in forming the bridged intermediate for an inner-sphere mechanism. Typical studies have involved cobalt(III), chromium(III), iron(II), ruthenium(II) and osmium(II) complexes. A typical example of a reaction involving two kinetically inert reactants is given in Eq. (9.25).

$$[Ru(bpy)_3]^{2+} + [Co(en)_3]^{3+} \rightleftharpoons [Ru(bpy)_3]^{3+} + [Co(en)_3]^{2+}$$

| $d^6$ | $d^6$ | $d^5$ | $d^7$ |
|-------|-------|-------|-------|
| inert | inert | labile | labile. |

$$(9.25)$$

In the case of other systems in which one or both of the reactants is labile, no such generalization can be made. The rates of these reactions are uninformative, and rate constants for outer-sphere reactions range from $10^{-9}$ to $10^{10}$ sec$^{-1}$. No information about mechanism is *directly* obtained from the rate constant or the rate equation. If the reaction involves two inert centers, and there is no evidence for the transfer of ligands in the redox reaction, it is probably an outer-sphere process.

However, *some* quantitative interpretation of the rates of outer-sphere reactions may be made. It is possible to determine the rate constant, $k_{12}$ for the reaction of two complex ions $[M^1L_6]^{2+}$ and $[M^2L_6]^{3+}$ (Eq. 9.26).

$$[M^1L_6]^{2+} + [M^2L_6]^{3+} \rightarrow [M^1L_6]^{3+} + [M^2L_6]^{2+} \qquad k_{12} \qquad (9.26)$$

Marcus and Hush have developed a theory, which bears their names, that relates the value of $k_{12}$ to the rates ($k_{11}$ and $k_{22}$) of the 'self-exchange' reactions of the two components (Eq. 9.27 and 9.28) and the stability constant $K_{12}$ for the reaction of interest.

$$[M^1L_6]^{2+} + [M^1L_6]^{3+} \rightarrow [M^1L_6]^{3+} + [M^1L_6]^{2+} \qquad k_{11} \qquad (9.27)$$

$$[M^2L_6]^{2+} + [M^2L_6]^{3+} \rightarrow [M^2L_6]^{3+} + [M^2L_6]^{2+} \qquad k_{22} \qquad (9.28)$$

In many cases, the values of $k_{11}$ and $k_{22}$ may be directly or indirectly determined. We shall say no more about this relationship here, other than to indicate that it proves to be generally applicable, and is sufficiently accepted that the Marcus-Hush equation is now used to establish when an outer-sphere pathway is operative. In the context of this chapter, the involvement of the $K_{12}$ term is interesting for it relates to the relative stabilization of various oxidation states by particular ligand sets. The factors which stabilize or destabilize particular oxidation states continue to play their roles in determining the value of $K_{12}$, and hence the rate of the electron transfer reaction.

There is a very special case for self-exchange reactions in which the left side of the equation is identical to the right side. Accordingly, there is no free energy change in the reaction, and the equilibrium constant $(K_{11})$ must be unity (Eq. 9.29).

$$[Co(NH_3)_6]^{3+} + [Co(NH_3)_6]^{2+} \rightleftharpoons [Co(NH_3)_6]^{2+} + [Co(NH_3)_6]^{3+} \qquad (9.29)$$
$$\Delta G^\ominus = 0 \quad \lg K = 0$$

However, metal ions in higher oxidation states are generally smaller than the same metal ion in lower oxidation states. In the above example, the $Co(\text{II})-N$ bonds are longer than $Co(\text{III})-N$ bonds. Consider what happens as the two reactants come together in their ground states and an outer-sphere electron transfer occurs. We expect the rate of electron transfer from one center to another to be very much faster than the rate of any nuclear motion. In other words, electron transfer is very much faster than any molecular vibrations, and the nuclei are essentially static during the electron transfer process (Fig. 9-6).

Thus, the interaction of the ground state cobalt(II) complex with long $Co-N$ bonds and the ground state cobalt(III) complex with shorter $Co-N$ bonds initially

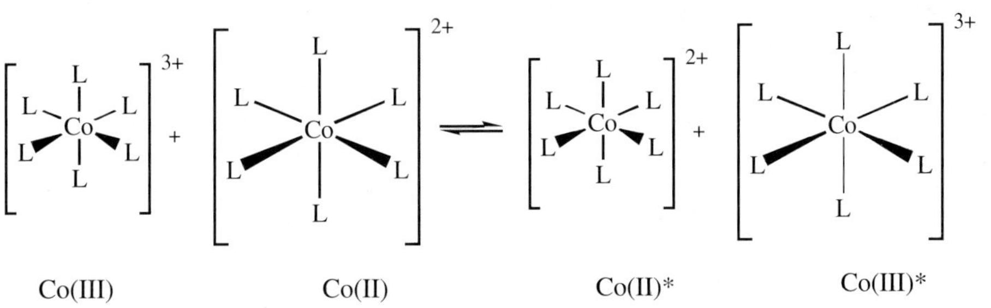

Co(III)          Co(II)          Co(II)*          Co(III)*

**Figure 9-6.** The consequences of a self-exchange electron transfer between a ground state cobalt(II) and a ground state cobalt(III) complex.

occurs without any rearrangement of Co−N bond lengths. Both products will, of course, be in vibrationally excited states. The cobalt(II) complex will have compressed Co−N bonds whilst the cobalt(III) complex will have extended Co−N bonds. At a later stage, these must relax to the equilibrium Co−N bond lengths appropriate for each oxidation state. This step involves the emission of energy. Yet there *is* no overall energy change in the reaction. The requisite balance derives from an activation energy associated with the electron transfer process. In order for there to be no overall energetic change in the electron transfer self-exchange reaction, the electron transfer must occur between vibrationally excited species with equivalent bond lengths (Fig. 9-7).

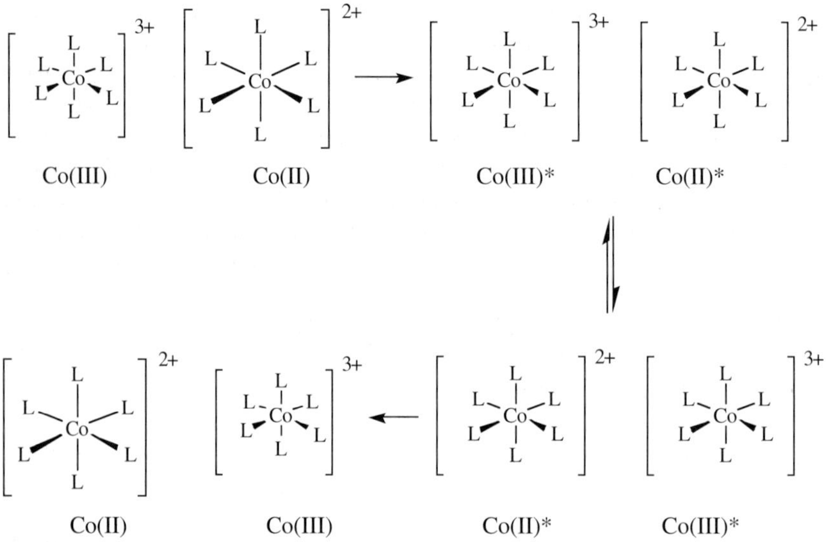

**Figure 9-7.** The self-exchange electron transfer reaction between vibrationally excited cobalt(II) and cobalt(III) complexes.

This is the origin of the various values for self-exchange rate constants. We may now attempt to rationalize some of these in terms of the *d*-electron configurations of the various oxidation states. Consider the self-exchange rate constants for some iron complexes.

| | |
|---|---|
| $[Fe(phen)_3]^{2+/3+}$ | $k = 10^8 \ M^{-1} \ s^{-1}$ |
| $[Fe(CN)_6]^{4-/3-}$ | $k = 10^3 \ M^{-1} \ s^{-1}$ |
| $[Fe(H_2O)_6]^{2+/3+}$ | $k = 10 \ M^{-1} \ s^{-1}$ |

The water complexes are high spin, whereas the cyanide and phen complexes are low spin. In the case of the cyanide and phen complexes, the interconversion of the $t_{2g}^6$ iron(II) and $t_{2g}^5$ iron(III) states simply involves the loss or gain of an electron from the $t_{2g}$ level. Since these are the orbitals oriented between the ligand donor atoms, there will only be minimal changes in the electron-ligand repulsions and small consequent changes in the Fe–ligand distances. The small changes in Fe–ligand distances mean that the activation energy for the electron transfer reaction will be low, and the rate of the reaction will be high. In the water complex, the $t_{2g}^4 e_g^2$ iron(II) and $t_{2g}^3 e_g^2$ iron(III) states are involved. Once again, the electronic changes are occurring in the $t_{2g}$ manifold. It is thought that the reason for the very rapid reactions of the phen complexes are due to the involvement of the $\pi^*$ orbitals of the ligand in the electron transfer process.

An interesting contrast is seen when we consider related reactions involving cobalt (Eq. 9.30). In this case, there is a spin state change in the electron transfer process. This results in two separate contributions to the high activation energy for the self-exchange. The high-spin cobalt(II) complex possesses two electrons in the $e_g$ orbitals. These are oriented directly towards the ligands, and electron-ligand interactions are expected to result in long Co–ligand distances. In these complexes, the Co(II)–N distance is 2.11 Å and the Co(III)–N distance is 1.93 Å. The activation energy is high because of this difference in bond lengths, but also because of the electronic rearrangement that is needed in the process. No longer do we simply move an electron from one center to another for now a rearrangement of electrons is to be achieved. For the same reasons underlying the Franck-Condon principle, we expect to have electron transfer between electronically excited states (Eqs. 9.31 or 9.32). Further discussion of this topic is beyond the scope of this book.

$$[Co(NH_3)_6]^{3+} + e^- \rightleftharpoons [Co(NH_3)_6]^{2+} \qquad k = 10^{-9} \text{ M}^{-1}\text{ s}^{-1}$$

$$\text{low spin} \qquad\qquad\qquad \text{high spin} \tag{9.30}$$

$$t_{2g}^6 \qquad\qquad\qquad\qquad t_{2g}^5 e_g^2$$

$$\text{Co(III)} \qquad \rightarrow \text{Co(III)}^* \qquad \rightarrow \text{Co(II)} \tag{9.31}$$

$$t_{2g}^6 \qquad\qquad t_{2g}^5 e_g^1 \qquad\qquad t_{2g}^5 e_g^2$$

$$\text{Co(III)} \qquad \rightarrow \text{Co(II)}^* \qquad \rightarrow \text{Co(II)} \tag{9.32}$$

$$t_{2g}^6 \qquad\qquad t_{2g}^6 e_g^1 \qquad\qquad t_{2g}^5 e_g^2$$

We conclude with a consideration of a few other cobalt self-exchange reactions. The reaction in Eq. (9.33) is faster than that involving the ammine complexes (Eq. 9.30) because the water is a weaker-field ligand than ammonia. Thus, the activation energy for the formation of the electronically excited states is lower, as is the change in Co–ligand distances in the two oxidation states.

$$[Co(H_2O)_6]^{3+} + e^- \rightleftharpoons [Co(H_2O)_6]^{2+} \qquad k = 1 \text{ M}^{-1}\text{ s}^{-1} \tag{9.33}$$

$$\text{low spin} \qquad\qquad\qquad \text{high spin}$$

$$t_{2g}^6 \qquad\qquad\qquad\qquad t_{2g}^5 e_g^2$$

The reaction in Eq. (9.34) is also faster because the bpy ligand is a strong field ligand and there is no longer any need for electronic rearrangement upon change in oxidation state. The process is now comparable to those discussed earlier for low spin iron complexes.

$$[Co(bpy)_3]^{3+} + e^- \rightleftharpoons [Co(bpy)_3]^{2+} \qquad\qquad k = 1\ M^{-1}\ s^{-1}$$
low spin                low spin                                                    (9.34)
$t_{2g}^6$                    $t_{2g}^6 e_g^1$

Finally, we consider the alternative mechanism for electron transfer reactions – the inner-sphere process in which a bridge is formed between the two metal centers. The $d$-electron configurations of the metal ions involved have a number of profound consequences for this reaction, both for the mechanism itself and for our investigation of the reaction. The key step involves the formation of a complex in which a ligand bridges the two metal centers involved in the redox process. For this to be a low energy process, at least one of the metal centers must be labile.

A number of ingenious experiments have been devised to establish the operation of this mechanism. These all revolve about the lability or inertness of particular $d$-electron configurations. Remember that complexes of first row transition-metal ions which possess $d^3$ or low-spin $d^6$ electronic configurations are usually particularly inert with respect to ligand substitution reactions – they are kinetically stabilized. Consider the reaction of the cobalt(III) complex $[Co(NH_3)_5Cl]^{2+}$ with the chromium(II) complex $[Cr(H_2O)_6]^{2+}$ in Eq. (9.35). The first step in an inner-sphere process would be the formation of a chloro-bridged complex. The chloride ligand is better suited for bridging than either of the neutral water or ammonia ligands. The $d^6$ cobalt(III) center is kinetically inert, but the $d^4$ chromium(II) complex is labile. The intermediate bridged complex is thus formed by the displacement of a water molecule from the chromium, and with retention of the cobalt–chloride bonding. If this complex collapses without electron transfer, it will be the labile Cr–Cl bond which breaks to regenerate the starting complexes.

$$[Co^{III}(NH_3)_5Cl]^{2+} + [Cr^{II}(H_2O)_6]^{2+} \rightleftharpoons [(NH_3)_5Co^{III}(\mu\text{-}Cl)Cr^{II}(H_2O)_5]^{4+}$$
$\quad d^6 \qquad\qquad\qquad d^4 \qquad\qquad\qquad\qquad d^6 \qquad\quad d^4$      (9.35)
$\quad$ inert $\qquad\qquad\qquad$ labile $\qquad\qquad\qquad\qquad$ inert $\qquad\quad$ labile

On the other hand, if electron transfer does occur within this bridged complex, a bridged cobalt(II)–chromium(III) complex is generated (Eq. 9.36). The $d^7$ cobalt(II) center is labile whilst the $d^3$ chromium(III) center is inert.

$$[(NH_3)_5Co^{III}(\mu\text{-}Cl)Cr^{II}(H_2O)_5]^{4+} \rightleftharpoons [(NH_3)_5Co^{II}(\mu\text{-}Cl)Cr^{III}(H_2O)_5]^{4+}$$
$\qquad\quad d^6 \qquad\quad d^4 \qquad\qquad\qquad\qquad d^7 \qquad\quad d^3$      (9.36)
$\qquad\quad$ inert $\qquad\quad$ labile $\qquad\qquad\qquad\qquad$ labile $\qquad$ inert

If this complex now collapses, it will be the labile Co–Cl bond which is broken, as opposed to the inert Cr–Cl bond. The labile cobalt(II) complex reacts further with bulk water to generate $[Co(H_2O)_6]^{2+}$ (Eq. 9.37). The key feature is that a necessary consequence of this inner-sphere reaction is the transfer of the bridging ligand from one center to the other. This is not a necessary consequence of all such reactions, but is a result of our choosing a pair of reactants which each change between inert and labile configurations. In the reaction described above, the chloride ion transfer does indeed occur, thus providing strong circumstantial evidence for the proposed mechanism.

$$[(NH_3)_5Co^{II}(\mu\text{-}Cl)Cr^{III}(H_2O)_5]^{4+} \rightleftharpoons [Co^{II}(NH_3)_5(H_2O)]^{2+} + [Cr^{III}(H_2O)_5Cl]^{2+}$$

$$
\begin{array}{llll}
d^7 & d^3 & d^7 & d^3 \qquad\qquad (9.37)\\
\text{labile} & \text{inert} & \text{labile} & \text{inert}
\end{array}
$$

## Suggestions for further reading

*The general textbooks mentioned earlier all discuss various aspects of the material presented in this chapter.*

1. E.C. Constable, *Metals and Ligand Reactivity*, Ellis Horwood, Chichester, **1990**
   – This discusses the metal-activated reactivity of ligands.
2. F. Basolo, R.G. Pearson, *Mechanisms of Inorganic Reactions*, Wiley, New York, **1967**
   – The classic text concerned with inorganic reaction mechanisms.

# 10 The Lanthanoid Series

## 10.1 The Lanthanoid Contraction

Thus far, we have focused exclusively upon the *d*-block metals. For some, the term 'transition elements' defines just these *d*-block species; for others, it includes the rare earth or lanthanoid elements, sometimes called the 'inner transition elements'. In this chapter, we compare the *d*-block and *f*-block (lanthanoid) elements with respect to their valence shells. In doing so, we shall underscore concepts which we have already detailed as well as identifying both differences and similarities between certain aspects of 'main' and 'inner' transition-metal chemistry. We make no attempt to review lanthanoid chemistry at large. Instead our point of departure is the most characteristic feature of lanthanoid chemistry: the +3 oxidation state.

The lanthanoids occur under scandium and yttrium in the periodic table. Some useful data are presented in Table 10-1. On crossing the series of +3 ions from lanthanum to lutetium, observe the variation of electronic configuration $4f^n$ from $n = 0$ to $n = 14$. Note too, a decrease in ionic radius by about 20% across the series. The reason for a size reduction across any series in the periodic table is, of course, the increase in effective nuclear charge that results from the inefficient mutual shielding of electrons in the same shell, together with the monotonic increase in real nuclear charge. However, the phenomenon is particularly marked in the lanthanoid series and is referred to as 'the lanthanoid contraction'. Perhaps the main reason for drawing attention to it by this special name lies in its consequences for the chemistry of the third row *d*-block elements that follow the lanthanoid series. That chemistry is strongly affected, as ever, by ionic radius. The increase in size of the third row *d*-block elements relative to those of the second row, which is expected from the increased number of electrons and the higher principal quantum number of the outer ones, is almost exactly offset by the intervening lanthanoid contraction. In consequence, there are more similarities between the chemistries of the second and third row transition metals than between the first and second row elements.

The magnitude of the contraction across the whole lanthanoid series is due partly to the length of the series – the placement of up to 14 electrons in the $4f$ shell – and partly to the poor shielding of one *f* electron for another. The poor shielding arises in two ways. Firstly, the form of the radial wavefunction for $4f$ electrons, like that of the $3d$, involves no inner maxima. Thus, these *f* orbitals are not of the 'penetrating' type, and so little inner electron density is available to shield the outer regions from the (increasing) nuclear charge. Secondly, *f* orbitals are often described as 'diffuse'. It is important to be clear about the use of this adjective here, however,

**Table 10-1.** Configurations and ionic radii for the lanthanoids.

| Element | Symbol | Electronic configurations[*] | | M³⁺ionic radius/Å |
| | | Atomic | M$^{3+}$ | |
|---|---|---|---|---|
| Lanthanum | La | $5d^1 6s^2$ | – | 1.061 |
| Cerium | Ce | $4f^2 6s^2$ | $4f^1$ | 1.034 |
| Praesodymium | Pr | $4f^3 6s^2$ | $4f^2$ | 1.013 |
| Neodymium | Nd | $4f^4 6s^2$ | $4f^3$ | 0.995 |
| Prometheum | Pm | $4f^5 6s^2$ | $4f^4$ | 0.979 |
| Samarium | Sm | $4f^6 6s^2$ | $4f^5$ | 0.964 |
| Europium | Eu | $4f^7 6s^2$ | $4f^6$ | 0.950 |
| Gadolinium | Gd | $4f^7 5d^1 6s^2$ | $4f^7$ | 0.938 |
| Terbium | Tb | $4f^9 6s^2$ | $4f^8$ | 0.923 |
| Dysprosium | Dy | $4f^{10} 6s^2$ | $4f^9$ | 0.908 |
| Holmium | Ho | $4f^{11} 6s^2$ | $4f^{10}$ | 0.894 |
| Erbium | Er | $4f^{12} 6s^2$ | $4f^{11}$ | 0.881 |
| Thulium | Tm | $4f^{13} 6s^2$ | $4f^{12}$ | 0.869 |
| Ytterbium | Yb | $4f^{14} 6s^2$ | $4f^{13}$ | 0.858 |
| Lutetium | Lu | $4f^{14} 5d^1 6s^2$ | $4f^{14}$ | 0.848 |

[*]Outside of closed [Xe] shell.

since they are not radially diffuse. Indeed, as we shall discuss at length shortly, they are radially compact. They are angularly diffuse in that $f$ orbitals have many more lobes than $d$ orbitals, for example. The $f$ orbitals have a larger angular spread and since they are normalized to unity – corresponding to their housing exactly one electron (of a given spin) – the local electron density is rather low. This, too, contributes to the poor self-shielding within the $4f$ shell.

## 10.2   The Core-Like Behaviour of $f$ Electrons

The radial spread of the $4f$ orbitals in lanthanoid +3 ions is so limited that $f$ electron density is almost totally confined to the inner regions of the lanthanoid ion. Figure 10-1 schematically compares the radial waveforms of the $4f$ and $6s$ orbitals. The situation is rather like that in Fig. 2-1 for the $3d$ and $4s$ orbitals in the first row $d$ block for higher oxidation state species. We have seen what consequences flow from the relative 'isolation' or 'uncoupling' of the $d$ electrons in such circumstances. In particular, recall how the chemical bonding between a $d$-block metal and its ligands is effected within a metal valence shell that is largely $4s$; and how the $d$ electrons modify that bonding by 'secondary', repulsive and other non-overlap

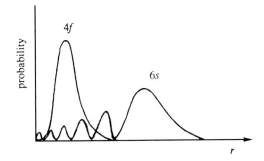

**Figure 10-1.** Radial probability functions for typical 4*f* and 6*s* orbitals.

means. The same is true for the lanthanoids, but to a greater extent. The core-like behaviour of the 4*f* orbitals in lanthanoid +3 ions is even more pronounced than that of the 3*d* orbitals in analogous *d*-block complexes. There is no significant contribution from the 4*f* shell to the valence shell in lanthanoid chemistry. This does not imply an ionic chemistry of $Ln^{3+}$ species, however, any more than that *d*-block bonding is primarily ionic. We discuss evidence of covalency in lanthanoid complexes below. First, we look at the consequences for the *f*-electron shell of its 'contracted' or core-like character.

## 10.3  Magnetic Properties in the *f* Block

The number of terms arising from a given $f^n$ configuration is generally much larger than from $d^n$ and follows directly from the greater degeneracy of the *f* shell. We shall see something of the complexity of *f*-block term diagrams in the next section. However, it is quite simple to work out the *ground* terms in the *f* block by using Hund's rules. For the $f^6$ configuration, for example, we maximize spin by placing each electron in a separate orbital (so reducing the interelectron repulsion energy).

| $m_l$ | 3 | 2 | 1 | 0 | -1 | -2 | -3 |
|---|---|---|---|---|---|---|---|
| | ↑ | ↑ | ↑ | ↑ | ↑ | ↑ | |

The total *z* component of the spin angular momentum, $M_s$, is given by the sum $\Sigma\, m_s = 3$ and implies a total spin for the ensemble of $S = 3$ and a spin-multiplicity $(2S + 1)$ of 7. Similarly, $\Sigma\, m_l = 3$, yielding $L = 3$. The ground term of $f^6$ is therefore $^7F$.

Spin-orbit coupling, which arises from the magnetic interactions amongst electrons, splits the $^7F$ *term* into *levels* $^7F_J$ with *J* values varying from the sum $(L + S)$ to the difference $|L - S|$, as usual. We find *J* ranging 0 to 6 in this case. For less-

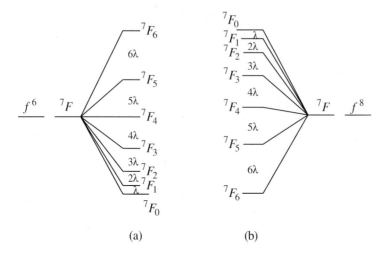

**Figure 10-2.** Splittings of $^7F$ ground terms for $f^6$ and $f^8$ configurations.

than-half-filled shells, as here, Hund's third rule places the level with the minimum $J$ value lowest in energy; so the ground level is $^7F_0$. Landé's interval rule defines the energy separation between two adjacent levels as $\lambda$, the spin-orbit coupling coefficient, times the larger of the two $J$ values. Collecting all these results together, we establish the splitting pattern for the ground term of $f^6$ as shown in Fig. 10-2a. It is left as an exercise for the reader to show that the corresponding ground term and level stacking for $f^8$ ions are as shown in Fig. 10-2b.

Now the magnitudes of the spin-orbit coupling coefficient are much greater in the $f$ block than the $d$. Consequently, the energy separations between levels for $f$-block ions are usually much larger than the ambient thermal energy, $kT$. For the $f^8$ ions of terbium(III), for example, $\lambda = 270$ cm$^{-1}$ and the first excited level, $^7F_5$, lies 1620 cm$^{-1}$ above ground. At room temperature, $kT \approx 200$ cm$^{-1}$ so that the population of the $^7F_5$ level is about $e^{-1620/200}$, which is negligible.

We saw in Chapter 5 that the paramagnetism of a system primarily depends upon the splitting of populated states within an applied magnetic field. So, for an (obviously unobtainable) sample of Tb$^{3+}$ ions, the magnetic moment is primarily a function of the $^7F_6$ level alone. There are second-order contributions arising from the mixing of higher levels into the ground level by the magnetic field but, as these are inversely proportional to the energy separation between the mixing levels, they are generally small. An analytical formula for the effective magnetic moment, $\mu_{eff}$, has been derived for the case where only one level, $^{2S+1}L_J$, is thermally populated and second-order contributions are ignored (Eq. 10.1).

$$\mu_{eff} = g \sqrt{J(J+1)} \qquad (10.1)$$

In Eq. (10.1), $g$ is the 'Landé splitting factor' and is given by the expression in Eq. (10.2).

$$g = 1 + \frac{J(J+1) - L(L+1) + S(S+1)}{2J(J+1)} \tag{10.2}$$

These formulae explicitly involve only the $L$, $S$ and $J$ quantum numbers that define the ground level $^{2S+1}L_J$, as required.

The discussion thus far refers to *free ions*. One can apply the formulae to the magnetism of $d$-block compounds but it fails utterly to reproduce experiment. The inapplicability of Eq. (10.1) is due to two factors: a) the smaller magnitude of $\lambda$ – at least in the first row $d$ metals – means that level splittings in the free ions are smaller and molecules significantly populate more than just the ground level, and b) the formula takes no account of the ligand-field splitting of the free-ion terms. We have seen that such splittings in the $d$ block are of the order of several thousand wavenumbers, a perturbation that dwarfs the effects of spin-orbit coupling. Indeed, spin-orbit coupling is manifest in the magnetic properties of first row $d$-block complexes largely as a correction to the 'spin-only' formula (though these corrections are very important for ions with $T$ ground terms).

The neglect of the ligand field in Eq. (10.1) leads one to expect no satisfactory account of the experimental magnetism of lanthanoid *complexes* either. It is an empirical fact, however, that Eq. (10.1) accounts extremely well for observed magnetic moments in most lanthanoid compounds. We compare typical experimental moments for lanthanoid complexes with those calculated from Eq. (10.1) in Fig. 10-3. Significant discrepancies occur for $f^5$ and $f^6$ species and we will comment on these shortly.

The question therefore arises of 'why does neglect of the ligand field, implicit in Eq. (10.1), not matter for the $f$ block while it is utterly unacceptable for the $d$ block?'. The answer is both trivial and subtle. Trivially, the neglect is acceptable

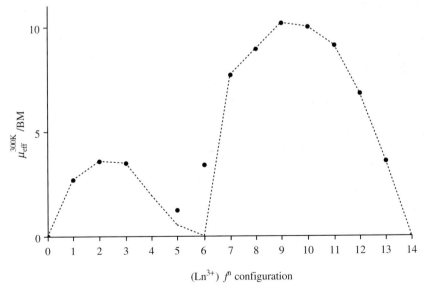

$(Ln^{3+})$ $f^n$ configuration

**Figure 10-3.** Comparisons between typical observed effective magnetic moments ($\bullet$) and those calculated (-----) with Eq. (10.1).

for the lanthanoids because the ligand field itself is negligible. The $f$ electrons are sufficiently buried beneath the valence shell so that they are affected very little by the ligand environment. The bond orbitals engaged in binding a lanthanoid metal to its ligands largely lie beyond the radial extent of the $f$ orbitals so that ligand-field splittings are very small. We shall see in the next section that splittings of tens of wavenumbers in the $f$ series replace the thousands of wavenumbers in the $d$ series.

The subtle reason, which we do not detail here, is as follows. The splittings caused by normal laboratory magnetic fields are of the order 0.1 to 1 cm$^{-1}$ only. These are, of course, small even compared with the small ligand-field splittings typical of lanthanoid complexes. One might expect, perhaps, that even these small ligand-field splittings cannot be ignored when considering magnetic properties. Actually, such expectations are not as reasonable as might first appear, partly because magnetism is about the *changes* that occur on application of an external field. Nevertheless, it requires a theorem due to Van Vleck to show generally and irrefutably that, so far as mean (spatially averaged) magnetic moments are concerned, ligand-field splittings which are no larger than about $kT$ have near negligible effects upon paramagnetism. Incidentally, this same theorem explains why so many of the simple formulae described in Chapter 4 work satisfactorily even when molecular geometries depart slightly from rigorous octahedral or tetrahedral symmetry. It is a crucial theorem for the theory of paramagnetism. This book, however, is not the place to demonstrate this important result.

Altogether, we can say that the success of Eq. (10.1) in reproducing the magnetic moments of lanthanoid complexes is due entirely to the very small magnitude of the ligand-field splittings and so, in turn, to the contracted nature of the $f$ orbitals.

We conclude this section with a further commentary on the discrepancies noted for the $f^5$ and $f^6$ systems. Consider first the case of the $f^5$ samarium(III) complexes. The ground term is $^6H$ with six levels ranging $J = 15/2$ to $5/2$. The $^6H_{5/2}$ level is lowest in energy with the first excited level lying $7\lambda/2$ above it, or some 1650 cm$^{-1}$. The second-order Zeeman effect, relating to the admixture of the first excited $^6H_{7/2}$ level into the ground $^6H_{5/2}$ level (and, indeed, of the yet higher lying levels) is not negligible in this case. More complete calculations which include these second-order effects, as Eq. (10.1) does not, do actually reproduce the observed moments for $f^5$ species very well. Second-order Zeeman terms are also important for $f^6$ europium(III) species. In this case, however, the ground level is $^7F_0$, as we showed above. Equation (10.1) yields a zero moment for this level. The same result can be arrived at as follows. The degeneracy of the $^7F_0$ level is $(2J+1) = 0$. A singly degenerate level cannot split in a magnetic field (or any other, of course) and so gives rise to no first-order paramagnetism. The first excited level, $^7F_1$, lies $1\lambda$ above ground, or about 230 cm$^{-1}$. (Note, by the way, that for $f^8$ ions, the $^7F_5$ level lies $6\lambda$ = 1380 cm$^{-1}$ above the ground $^7F_6$, so illustrating a major difference between $f^n$ configurations and their 'hole equivalents'). Accordingly, there is a significant population ($e^{-230/200}$) of the first excited level for $f^6$. So, in addition to any second-order Zeeman corrections, we must include first-order terms relating to electrons populating this excited level. Once more, a full calculation of these effects does indeed reproduce the magnetic moments that are typically observed for $f^6$ species.

Overall, then, the magnetic moments of all lanthanoid complexes are well reproduced without reference to the ligand field; *inter alia*, we can infer that the ligand-field splittings in *f*-block complexes are no greater than about *kT* at room temperature.

## 10.4 Spectral Features

Part of the absorption spectrum of an aqueous solution of neodymium(III) – configuration $f^3$ - is shown in Fig. 10-4. The situation shown there is quite typical of the whole of the lanthanoid series *i.e.* we could have chosen any $f^n$ configuration equally well to illustrate the main characteristics of the spectra of lanthanoid complexes. We shall focus on three main features: splittings, band widths and absolute excitation frequencies.

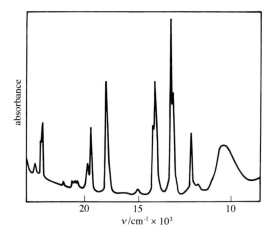

**Figure 10-4.** Absorption spectrum of an aqueous solution of $Nd^{3+}$ ions.

The solution spectrum is remarkably similar to that of the corresponding free ion in that slightly split groups of transitions replace the free-ion spectral lines on a one-to-one basis. These splittings are the ligand-field components. That they occasionally seem so numerous is due to the more complex geometry of the species, which is probably $[Nd(H_2O)_9]^{3+}$, as well as to the fact that the greater degeneracy of the *f* shell relative to the *d* shell begets more sublevels even for the same coordination geometry. We shall not concern ourselves with the details of these splitting patterns. Suffice to note that here is the direct evidence of the very small ligand-field perturbations that were deduced from the magnetic properties discussed in Section 10.3.

By and large, the spectral bands are very sharp as compared with '$d-d$' transitions in $d$-block complexes. The sharpness of these '$f-f$' transitions follows immediately from the core-like character of the $f$ shell. It interacts little with the bond orbitals and other aspects of the environment – hence the small ligand-field strength – and so the range of interaction throughout a molecular vibration is also small. The intensities of these transitions are also small. Typical extinction coefficients for '$d-d$' transitions in centrosymmetric complexes are of the order 5–10. For acentric chromophores, like the tetrahedra, they may be around 500. Those in this (typical) lanthanoid system are about 5. Considering that the coordination is probably that of a tricapped trigonal prism, which is non-centrosymmetric, the '$f-f$' intensities are some two orders of magnitude less than might be expected for a similar $d$-electron system. Once again, we understand this result in terms of the much smaller overlap between the metal $4f$ orbitals and the bond orbitals. The metal $f$ – ligand orbital mixing cannot be zero, for otherwise Laporte's rule would ensure vanishing intensities, but it is very small: smaller than for the $d$ block and, as we saw in Chapter 4, that is small anyway.

Before moving on to the absolute transition energies in lanthanoid spectra, let us take stock. The resemblance between the transition energies of lanthanoid complex spectra and those of the corresponding free ions, taken together with the sharpness and weakness of the bands and the small magnitudes of the ligand-field splittings, all concur with the notion of a well-buried $f$ shell. The magnetic moments of lanthanoid complexes similarly support this view. The $f$ electrons comprise a well decoupled subset of electrons within these complexes, bequeathing to the metal $6s$ (and perhaps other) orbitals the role of the valence shell. We thus observe a situation like that described for the $d$ electrons in the main transition-block (in higher oxidation states) complexes, but much more obviously. Both classes of compounds, however, are covalent in that complex species retain their integrity in many environments.

Evidence for that covalency comes directly from our last topic, namely, the absolute transition energies of complex '$f-f$' spectra. Many of the spectral bands in the spectra of lanthanoid complexes involve transitions between levels of the same term and, as such, provide a measure of the strength of the spin-orbit coupling. It is generally observed that these interlevel spacings are smaller than in the spectra of the corresponding free ions. Thus, $\lambda$(complex) < $\lambda$(free ion). Other transitions occur between components of different terms. After appropriate (and, unfortunately, rather complicated) analysis, one may determine the magnitudes of the various interelectron repulsion parameters, which include, for example, the Racah $B$ parameter discussed in Chapter 6. It is found, quite generally, that $B$(complex) < $B$(free ion). The magnitudes of these nephelauxetic effects are roughly of the *same* order as found for $d$-block complexes. Similarly, the reductions in $\lambda$ values, as above, – and obtained again only after lengthy analysis – are of similar proportionate magnitudes in $d$- and $f$-block systems.

Both phenomena attest to the covalency of the chemical bonding in these species. Incidentally, they also highlight the different characters and implications of the spectrochemical and nephelauxetic series. Within either lanthanoid- or (higher oxidation state) $d$-block species, the ligand orbitals overlap with the metal $s$ functions and donate electron density to the metal. The $s$ orbitals ($4s$ for the first row $d$ block; $6s$ for the lanthanoids) are of the penetrating type (inner maxima) and that portion of the ligand electron density which occupies these inner regions is particularly effective in shielding the outer, 'spectral', $4f$ or $3d$ electrons from the nucleus. The $f$ (or $d$) orbitals expand somewhat and so the average distance between $f$ $(d)$ electrons increases and the interelectron repulsion parameters decrease. It is also the case, though we do not enlarge on the matter here, that the magnitudes of spin-orbit coupling coefficients are inversely related to a power of the mean distance of (spectral) electrons from the nucleus. The reduction in $\lambda$ values goes hand-in-hand with the reduction in $B$ values, though not *pro rata*.

The central point, then, is that tiny ligand-field splittings and 'normal' sized nephelauxetic effects in lanthanoid spectra are not at all contradictory. The one reveals the isolation of the $f$ shell, the other attests to the normality of the metal– ligand bonding.

# Suggestions for further reading

1. *Systematics and Properties of the Lanthanides* (Ed.: S.P. Sinha), Reidel, **1983**.
   – Here, the article by Hüfner shows energy levels throughout the $f$-block.
2. J.H. van Vleck, *The Theory of Electric and Magnetic Susceptibilities*, Oxford University Press, Oxford, **1932.**
   – This is a great original – see Chapter 9.
3. S.A. Cotton, *Lanthanides and Actinides,* MacMillan, Basingstoke, **1991.**

# Index